Architectural Terra Cotta

Architectural Terra Cotta examines the evolution of terra cotta and prepares architects and builders to make new, creative uses of the timeless material. Terra cotta is among the oldest of manufactured building products, yet it has once again become a material of choice in contemporary façade design. From the walls of Babylon to high performance rainscreens, terra cotta claddings have repeatedly proven to be technically superior and aesthetically triumphant. Understanding the evolution of terra cotta prepares architects to add new, creative chapters to a rich history.

This book describes the key attributes that recommend the use of terra cotta and explain its continuing success. The core of the book traces the many ways that terra cotta can be formed, finished and applied to buildings. These techniques demonstrate the full potential of the material, showing how its unique capabilities have been developed over time. A comprehensive inventory of recent examples, project case studies and architectural details, this book provides a basis for understanding the nature of the material and the opportunities it offers in new work.

With over 150 color images, this volume provides a concise resource for all those considering terra cotta as a façade system: architects, façade engineers, cladding subcontractors, materials suppliers, developers and prospective clients. With inspiring examples of expressive possibility, this invaluable book will find a home with students and professionals alike interested in making rich, colorful and durable buildings.

Donald B. Corner is a practicing architect and member of the College of Distinguished Professors of the Association of Collegiate Schools of Architecture. He has been recognized for research and innovative teaching in building construction and detailing.

John Rowell is a professor of architecture and founding principal of Rowell Brokaw Architects, an award winning firm in Eugene, Oregon, USA. His practice, research and teaching focus on sustainable, high performance building solutions at a variety of scales.

For twenty-five years, Professors Corner and Rowell have offered an integrated course in the technology and design of building enclosure at the University of Oregon.

Architectural Terra Cotta

Donald B. Corner and John Rowell

Routledge
Taylor & Francis Group

NEW YORK AND LONDON

Cover image: Banca Popolare di Lodi, Italy. Photo: Donald Corner.

First published 2022
by Routledge
605 Third Avenue, New York, NY 10158

and by Routledge
4 Park Square, Milton Park, Abingdon, Oxon, OX14 4RN

Routledge is an imprint of the Taylor & Francis Group, an informa business

© 2022 Donald B. Corner and John Rowell

Library of Congress Cataloging-in-Publication Data
A catalog record for this title has been requested

ISBN: 9780367178260 (hbk)
ISBN: 9780367178307 (pbk)
ISBN: 9780429057915 (ebk)

DOI: 10.4324/9780429057915

Typeset in Minion Pro
by codeMantra

Printed in the UK by Severn, Gloucester on responsibly sourced paper

Contents

Preface

The Lombardy region of Italy, south and east of Milan, is home to a collection of fourteenth- and fifteenth-century buildings that make striking use of terra cotta as a façade material. The refined, sculptural role of clay on display evolved in a culture that is rich with brick and carved brick antecedents. In 1867, Lewis Gruner edited an account of these buildings. Team members Vittore Ottolini and Federigo Lose analyzed and recorded the façades, declaring them to be in remarkable condition. At the Cathedral of Crema, "the fine material (terra cotta) remains to the present day in excellent preservation, and the angles are as sharp as if newly cut."[1] This was an important endorsement as architects of that period were seeking an expressive material that could withstand environmental attack in the changing atmosphere of the new industrial city. Applications of terra cotta proliferated, as it was easy to clean and more durable than stone.

In the same region, in 2001, Renzo Piano covered the façades of the Banca Popolare di Lodi with a beautifully articulated system of terra cotta panels. He worked with producers from Il Ferrone, a village in the hills south of Florence, source of a classic red terra cotta well known since the roof tiles of Brunelleschi's dome. Piano's approach evolved out of earlier experiments in brick, as he developed a system of pre-assembled panels to produce a contemporary rainscreen wall. Another dramatic expansion in terra cotta use followed, as architects were searching for a way to recapture the aesthetics of small clay units while clearly freeing the material from any past associations with the bearing of structural loads.

These were two very different project teams, each searching for a building material that would meet challenges they had identified in their era. They each rediscovered the virtues of terra cotta. It is a truly resilient technology, one that has been adapted to changing circumstances across the full span of construction history. Each team capitalized on known solutions to recurring problems, and each contributed in their own way to a new surge of creative achievement in the material.

[1] Gruner, notes to plates 16–22.

Deeply meaningful buildings, those that resonate with a particular time or place, are grounded by the skillful use of basic materials: wood, stone, metals and burned clay. These materials continue to intrigue and inspire because they retain timeless, elemental qualities even as their use has grown more sophisticated in both technical and aesthetic terms. This volume is devoted to tracing the unique qualities that have recommended terra cotta from the ancient past to the present day.

Acknowledgments

The authors would like to extend a special thanks to Hannah Zalusky, a consummate professional without whom this project would not have been possible. Hannah prepared all of the illustrations and processed the photographs that are the heart of this book. She did this while excelling in her graduate studies at the University of Oregon.

Our colleagues from the terra cotta industry gave freely of their time, hosting us for plant and office tours, and following up with answers to our many questions: John Krouse and his staff at Boston Valley Terra Cotta, Jamie Farnham at Gladding McBean Company, Christian Lehmann and Bud Streff at NBK North America, Daniela Wienen and Stefan Verriet at NBK Keramik, and particularly Alessandro Piazza who got things started with repeated welcomes to Palagio Engineering in Italy.

We recognize the unique contributions of Maurits van der Staay, through writings and drawings that appear in the case studies at the end of the book. They reflect his passion for detail and productive association with Renzo Piano Building Workshop.

Throughout the volume, we acknowledge the essential contributions of the many architects, technical consultants and subcontractors who collaborated on projects that demonstrate the extraordinary potential of terra cotta in creative hands. They have provided drawings, photographs and descriptive materials to the benefit of students and emerging professionals. With us, they have shared a commitment to education and the future of our chosen fields.

A profound thanks to Professor Jenny Young, critic and companion over years of architectural travels, and first reader of this text.

Introduction

This book is for colleagues who share a curiosity about buildings, with an appreciation of those we have inherited from the past, and the ambition to enrich those we will make in the future. The focus of the book is the making of building enclosures: the structured pattern of solids and voids through which we control the passage of light, heat, water and air. The visible, outward surfaces are façades that collectively form the walls of our shared, civic spaces. Façades must resist extremes of exposure that will work on them over time. In searching for materials to meet these performance goals, there are basic questions to answer:

Why choose this material?
What is it capable of doing?
Where and how is it best applied?

This volume offers a portrait of terra cotta, exploring the breadth of possibility it offers in response to these questions. The chapters identify key attributes of terra cotta as a material, a product and a system of enclosure. Opening with a brief survey of historical precedents, the chapters trace the many ways that terra cotta has been formed, finished and applied to buildings over time. The body of the book is a structured inventory of contemporary methods, including example projects and construction details. Together they frame the creative range of the material and reveal the continuing influence of fundamental principles on forms of the present day.

This collection is offered as a resource for the ongoing study of architecture, by those in school and in practice. The structure reflects a teaching tradition installed at the University of Oregon by Associate Professor Michael Shellenbarger, an early mentor to the authors. Trained in strict

DOI: 10.4324/9780429057915-1

modernism, Mike developed over his teaching career a greater and greater appreciation for the richness of historic building methods, particularly terra cotta. Courses at Oregon trace fundamental materials from their sources, through all of the processes that shape them, to establish a basis for design. Historical examples illuminate reasons the material was preferred at different moments in time, allowing us to understand what has changed and what remains the same; what is rediscovered, and what is truly new. This volume is a snapshot taken from the Building Enclosure course team taught by the authors for twenty-five years. The extended example of this one material is used to demonstrate an overall approach to the recurring principles of building enclosure, as the science continues to evolve.

Designers of the weather envelope must engage the full spectrum of sustainable building techniques. At one end are low impact materials, like wood shingles, that can be periodically replaced. At the other end are high first cost materials that must be amortized over a long service life. The examples chosen here place terra cotta within this continuum. It is a long-term material, and depending on the strategies employed, it can be very competitive on a life cost basis. Longevity in the envelope requires more than just satisfying the technical requirements of durability. To return their maximum benefit, façade materials must be maintained in place by owners who find them visually and emotionally satisfying. The objective of this book is to demonstrate that terra cotta is such a material because it has capacities that inspire designers to make fresh, new uses of it, each in their own time.

CHAPTER 1

Origins

The origins of architectural terra cotta emerge from the history of earthen construction, as traced by Norman Davey in *A History of Building Materials*. Babylonians and Assyrians applied burned clay as cladding, using wall sections in which the weathering surface was consciously differentiated from the body of the building underneath. Beyond the functional advantages, the distinct outer shell provided a place to present symbols, narratives and ornament.

> The famous Isthar gate in Babylon, built by Nebuchadnezzer II (604–562 BCE) … was faced with kiln baked bricks bedded in bitumen.[1]
>
> Each large city usually had its tower, or ziggurat, on which a shrine was erected, and these were of sun baked brick cased in burnt brick, but one at the tower of Erech had its outer surface protected by thousands of pieces of pottery hammered into the brickwork while it was slightly plastic.[2]
>
> Both the Babylonians and the Assyrians from the ninth to the sixth century [BCE] made patterned bricks and wall tiles with colored glazes, and the Palace of the Achaemenid Kings of Persia at Susa had brick friezes decorated in relief in this manner. A particularly fine example is the "Frieze of the Archers" from Susa, dating from about 500 [BCE], depicting figures of the royal bodyguard, now in the Louvre, Paris.[3]

Architectural terra cotta has physical and functional antecedents in ancient times, yet the name itself is surprisingly recent. "Since Antiquity, builders have used fired clay for architectural features, but the term 'terra cotta' (baked earth) dates from the eighteenth-century revival of the

[1] Davey, p. 67.
[2] Davey, pp. 23–4.
[3] Davey, p. 69.

DOI: 10.4324/9780429057915-2

medium."[4] In 1886, at the apex of that revival, James Doulton defined architectural terra cotta as follows:

> That class of ware used in the construction of buildings which is more or less ornamental and of a higher class than ordinary bricks, demanding more care in the choice and manipulation of the clay and much harder firing, and being, consequently, more durable and better fitted for moulded and modelled work.[5]

Terra cotta is distinguished from other clay components through its architectural application, more than purely technical differences. At the time of Doulton's definition there were surely tall, industrial chimneys built of bricks with greater compressive strengths than those needed for claddings.[6]

Terra cotta is a process as well as a product. It is defined by a combination of factors that only begins with the clay. Operational definitions of clay describe naturally occurring, ultra-fine soils that are plastic and cohesive when damp, leather hard when dried, and a brittle but rock-like mass when fired. A summary of clay origins, molecular structure and related behaviors can be found in *Earth, Brick and Terra Cotta*, edited by members of Historic England's Building Conservation and Research Team.[7]

> Clays are created by chemical decomposition of certain bedrock minerals due to the actions of weathering and hydrothermal and biological processes, and physical disintegration. … Some deposits remain close to the bedrock from which they formed (primary or residual deposits). Others have been carried considerable distances by water, wind or glacial action (secondary or transported deposits). In the course of transportation, soil particles may become segregated according to size, and be deposited in separate layers or beds, as gravel (larger than 2mm); sand (2mm - 0.063mm); silt (0.063mm – 0.002mm); or clay (less than 0.002mm).
>
> Clay minerals are inorganic crystalline substances with an atomic structure consisting of sheets of silica and alumina (or sometimes magnesia) arranged in parallel layers. The stacking arrangement of these layers, and the ions and water molecules that link them, define the various clay minerals, and influence their respective properties and behavior.
>
> Clay-rich soils remain plastic and deformable over a wide range of moisture contents. The upper and lower limits of this behavior are defined as the liquid limit and plastic limit respectively. At moisture contents above the liquid limit, clay particles are suspended in free water

[4] Simpson, p. 129.
[5] As quoted in Stratton, p. 13.
[6] Lehmann, 2020.
[7] Henry et al.

and the mixture will flow. But when the moisture content falls below the plastic limit, the particles bind together, and the material becomes brittle.[8]

Brick, through most of history, has been known as a regional material. Clay of sufficient quality was found close to the point of end use, giving bricks distinctive local characteristics, particularly color. Terra cotta companies also established plants at particular clay sources. A cluster of villages south of Florence, Italy, referred to as Impruneta, are known for local clay that delivers products of exceptional frost resistance and durability, ranging from garden pots to architectural components.

In the eastern United States, a consensus emerged that it was less expensive to ship raw material in bulk than it was carefully packed finished pieces. The markets in Chicago and New York were large enough to sustain production facilities close to the construction sites. Whereas forty-eight major American firms once produced terra cotta for building façades, there are now only two remaining: Gladding McBean in Lincoln, California and Boston Valley in Orchard Park, New York. Gladding McBean was established in a rural area north of Sacramento where a bank of pure white kaolin clay had been discovered by accident while local officials were straightening a road. The founders built a large facility that shipped products up and down the West Coast. Boston Valley is supplied with clay from Ohio, long a quality source in the United States. Finished products are shipped far and wide, but significant demand comes from New York City. Terra cotta uses higher grades of clay, in far less quantity than brick production. It is also a higher value end product, allowing the shipping costs to be more readily absorbed into the price.

Contemporary architectural terra cotta moves about within an international marketplace. Top quality, custom work comes from a limited number of production sites. The site chosen or assigned to fill a particular order has to do with the skills and capacities at that plant more than the properties of local clay. The needs of a factory might be filled with clays from a number of suppliers, both near and far.

Clay is mined, usually in open pits. Traditionally, it was stockpiled and allowed to weather for up to a year before use. This was understood to break down the clay lumps in size and allow chemical conversions to reduce the level of common impurities. Modern-day suppliers use mechanical processes to prepare clay without extended weathering.[9]

Clay suppliers may prepare materials for both brick and terra cotta producers, using different techniques. In wet preparation, roller grinders crush the clay and impurities to a specified size. Water is added during

[8] Henry et al., pp. 6–7.
[9] Prudon, p. 63.

the process and the clays are stored for a few weeks so that the moisture levels are evenly distributed. In dry preparation, mechanical grinding occurs without water. Hot air is added and carries the finely ground particles away into storage. Grinding principally adjusts the particle size of the aggregates and impurities in the mix, since the clay is already very fine. Impurities must be small enough to have no negative effects on the end product.[10]

There are three major clay types used for terra cotta: china clay (pure kaolinite), ball clay (kaolinite with quartz and mica), marls and fireclay (more quartz than kaolinite, plus other metal oxides and impurities like iron).[11] These differ as to plasticity, shrinkage and firing temperature. The history of terra cotta companies records a range of preferences. In ceramics as a craft, clay types have great importance to the artisan. For commercial production of architectural components, the selection criteria are much more pragmatic. Clays must deliver adequate structural strength at reasonable firing temperatures, since energy is a huge component of production cost. Low water absorption and limited impurities are important. Beyond these, the best clay is the one best suited to the processes that will be used by a given company. There must be a secure and consistent supply so that recipes can be reproduced reliably over time. This relates to both appearance and performance. If a large-scale project is built in phases, it must be possible to duplicate the outcome several years later. As a leading producer, NBK Architectural Terra Cotta has plants in Germany, Portugal and China that can all run the same clay. In scheduling production, it may be advantageous to assign various components of a large project to different plants.[12]

Clay enters the production process as bulk dry powders. Mixtures can then be determined by weight for better quality control. Clay alone develops high strengths, but with shrinkage of 12–14%. At this level, formed pieces would be hard to dry and fire without tension cracks. Instead, clay is mixed with "grog," which is previously fired scrap material that has been ground to particles 1.5 mm (0.06 in) in diameter. Mixing in 30–35% grog reduces shrinkage to 7–8%.[13]

The ratios of clays, grog and additives for color change define a particular recipe, coupled with the amount of water to be used. Computer control of the quantities reproduces clay batches with the workability and shrinkage rate that has been anticipated in the design of a component. Without computer control, test samples from a batch have to be dried and fired to confirm the shrinkage. Heavy duty mixers can produce a ton of plastic material in a batch. In the past, mixing might have taken an hour, but now

[10] Lehmann, 2020.
[11] Henry et al., p. 640.
[12] Lehmann, 2020.
[13] Lehmann, 2020.

it can be completed in 4–5 minutes. The results are ready to move on to the forming process.

Once clay products are formed, they must be dried carefully. While that process is not technically complex, the pattern and rate of drying controls shrinkage movements. Drying of large pieces is hardest to control. Since they dry from the edge to the center, shrinkage will introduce tension stresses near the perimeter at a time when the green clay has little strength to resist cracking.

The firing process has distinct phases that begin with evaporation of free water from the pores, resulting in more shrinkage. At a higher temperature level, carbonaceous materials are burned out if there is oxygen available to complete the process.

> When clay is heated to a temperature above 600°C it undergoes irreversible "ceramic change" when water which forms part of the crystal structure is driven off. … At temperatures above 800°C, the strength and durability of the clay body increase progressively as particles become welded together with glassy material ("vitrification") and new minerals form and recrystallize. Important among these changes, for the development of mechanical strength, is the formation of the mineral "mullite", which begins at temperatures between 950°C and 1050°C.[14]

The temperature ranges quoted depend on the composition of the materials. Secondary minerals present or added to the clay act as fluxes to influence the temperature at which vitrification takes place. They also affect hardness, porosity, color and shrinkage.

Rather precisely at 573°C (1063°F), "quartz inversion" occurs, with the crystal structure of fine quartz particles changing in form. On the heating side of the cycle, this inversion increases the volume of the silica, but the change is absorbed by the clay body. During cooling, there is a sudden reduction in volume at the same temperature, which can result in tension cracking of the now brittle material. It is important to cool slowly through this transition to prevent defects in the product.[15]

Numerous quality control checks are conducted as fired elements emerge from the kiln. Extruded planks are tested for trueness with a straightedge that matches the designed length of the final product. Tiles supported at the ends are loaded to confirm bending strength for wind resistance. Spraying the tiles with water will reveal any hairline cracks. The most intriguing test is to tap a tile with a hammer. Acceptable units will ring like a bell, whereas flawed units will resound with a clunk.

[14] Henry et al., p. 9.
[15] Lehmann, 2020.

CHAPTER 2

Durability

Proponents of a new or rediscovered material often forecast extraordinary performance. In 1884, Charles Thomas Davis wrote:

> In faithfully made and vitrified terra-cotta, we have the great and only lasting triumph of man over natural productions; for timber will rot, stone, even granite, will disintegrate, iron will oxidize, these and all other metals will succumb to the action of fire, and other destroying influences of the elements; but properly made and thoroughly burned terra-cotta will pass through the centuries, and be the last to yield to those influences to which all natural productions must give way, the material not only being absolutely fireproof, but also in all architectural employments practically time proof and indestructible.[1]

A measure of skepticism is due to assessments arising from within a particular industry. Early proponents of reinforced concrete also proclaimed invincibility. Neither foresaw damage to their assemblies from expansive corrosion of ferrous metal components. Ultimately, terra cotta is a manufactured material. Long-term performance is developed only through the skills and experience of the producer. When "faithfully made and vitrified" the durability of terra cotta is extraordinary. Pottery is found in the excavated remains of every ancient culture that had access to clay and fuel to fire it.

The history of burned clay in architecture tracks the history of kilns. As durability is dependent on the ultimate firing temperature, the development of the double chamber, updraft kiln raised performance expectations by reaching temperatures of 1000°C (1832°F) or higher. Evidence of such infrastructure remains wherever there was a stable society. Norman Davey compares excavations from Khafaje, Diyala Province, Iraq (third millennium BCE), with those of Roman sites from two thousand years

[1] C.T. Davis, p. 296.

DOI: 10.4324/9780429057915-3

later.[2] Terra cotta fragments found all across the Persian, Greek and Roman worlds indicate advanced production techniques with both the quantity and quality to impact building practices.

The confidence that ancient cultures held for terra cotta is demonstrated by where it was first applied. Shelter begins with a system for shedding rain, and overhanging roofs extend their protection to walls of lesser construction. The roof is the most demanding exposure, with high temperature gains from direct sun and dramatic radiant cooling to the night sky. In Babylonia, as well as Greek and Roman sites, the first application of burned clay products was in the form of roof tiles and their companion parts.[3] This continued through Medieval architecture when the limited amount of terra cotta used was devoted to roofs and floors.[4]

The resilience of fired clay in wet locations was demonstrated by its use in foundations, damp proof courses under mud brick walls, and particularly in pipes. Ancient cultures with limited access to vitrified clay products conserved them for these extreme exposures until larger and better kilns made them available to more general uses.[5] Terra cotta pipes are still prized for their long-term resistance to a range of fluids considerably more aggressive than rain water. Innovative nineteenth-century adopters of architectural terra cotta in the United States frequently relied upon pipe factories to produce their ornamental components.[6] As competitive suppliers of refined terra cotta façades grew in number, many were sustained through cycles in the architectural market by their reliable sales of pragmatic, durable commodities. Gladding McBean of California still produces large diameter pipe.

Terra cotta in building has a history of use intertwined with that of stone. Both were considered durable, with the former more accessible and the latter more exalted. The roof tile migrated downward on early Greek temples to become a protective shell for the supporting brick and wood construction. Ornamental details were first executed in terra cotta because it was so easy to form. As the Greek culture flourished, stone replaced the pragmatic elements of clay.[7] Etruscan temples made similar uses of terra cotta: at the roof, the cornice, and as explicit covers over wood beams and other vulnerable locations lower down the building. These structures were described by Vitruvius and a reconstruction can be seen in the courtyard of the National Etruscan Museum at the Villa Julia in Rome. The Etruscans did not make a transition to stone, despite the widely acknowledged influence of Greek temple design on their own.

The "Frontone di Talamone" is a magnificent bas relief from the pediment of an Etruscan temple, dated at 150 BCE.[8] The terra cotta figures were found late in the nineteenth century on a hill overlooking the bay

[2] Davey, p. 66.
[3] Davey, p. 69.
[4] Elliot, p. 52.
[5] Davey, p. 67.
[6] Tunick, p. 7.
[7] Elliot, p. 52.
[8] Hall.

at Talamone in southwestern Tuscany. In their own way, they evoke the grandeur of the marble sculptures on the Parthenon (447–431 BCE), famously removed by Lord Elgin to the British Museum. The Capitoline Museums in Rome display fragments from the pediment of a Republican era temple that once stood between the Palatine and Caelian hills. The terra cotta cornice and figures also date from the middle of the second century, BCE. Later, Imperial Rome prospered, and like the Greeks, patrons of architecture called for marble.

Renaissance architecture derived from classical precedents in stone. During his training, Andrea Palladio (1508–80) traveled to Rome to measure and draw all of the monuments. Nevertheless, in his own work he limited the use of carved stone because it was very expensive. He built column shafts and rusticated walls of brick that were often but not always covered in lime plaster. As Palladian influence migrated to England, there was undoubtedly an appreciation for the economy and nobility of his construction language. By the early eighteenth century there was a conscientious search for alternative methods to produce artificial stone, with experimentation in fine cements.[9] Ultimately, the answer was a return to terra cotta.

In 1722, Richard Holt and the architect Thomas Ripley took out a patent for "A certain Compound Liquid Metal" from which artificial stone could be made, formed in molds and vitrified by "Strong Fire."[10] The latter references implied that it was a ceramic product, although there was a great deal of intrigue associated with their "secret" formula. By 1732 a confidential affidavit identified the distinctive ingredient as ground glass, mixed with the clay. Holt published a long list of components available from his factory, but his business did not survive a building slump in London during the 1730s and 1740s.

In 1769, the Eleanor Coades, mother and daughter, bought into an obscure artificial stone company located in Lambeth, now a borough of South London. Ultimately, the company was able to supply architectural elements of quality, durability and accurate dimensions. Another advantage over profiles run in plaster was the ability to prefabricate pieces off-site, reducing construction time. The product was originally named "Lythodipyra," which in Greek meant twice fired stone. It gained market acceptance as "Coade Stone."

The formula for Coade Stone was still somewhat a secret, inherited through a chain of individuals beginning with Holt. By 1818, it was more or less known as a "species of terra cotta." Finally analyzed with electron microscopes in the late 1980s, it contained: 50–60% ball clay from Southwest England, 10% grog, 5–10% crushed flint, 5–10% fine quartz or sand and 10% crushed lime soda glass. These constituents gave it a gritty texture

[9] Stanford, p. 97.
[10] Quoted in Stanford, p. 97.

that was closer to natural stone than other ceramics. The glass apparently came from the waste stream of a bottle factory in the same village.[11]

The success of Coade Stone was one of marketing more than formula. It was positioned as superior to natural stone because it was resistant to frost and retained sharp detail when used as a park sculpture or monument. The success was also built on a rich catalog of parts that could be mixed and matched. Whether or not more durable than all forms of stone, Coade Stone was certainly superior to molded stucco, which for reasons of first cost was its closest competitor. Despite its remarkable properties, Coade Stone was so strongly identified with late Georgian architecture of the period that they died out together in the 1830s when fashion changed.[12]

In the succeeding Victorian age, terra cotta had different sources: traditional pottery producers, and then companies that grew up where both clay and coal were available. A seminal building was the St. Stephen and All Martyrs Church at Lever Bridge, Bolton (1842–5). It was entirely terra cotta, inside and out, including extensive Gothic details that critics said should be exclusively reserved for stone.[13] The advantages of durability and cost prevailed.

> Manufacturers made great claims for the benefits of terracotta as a building material. It was argued to be more durable (weather- and fire-resistant) and easier to maintain than stone or brick, and resistant to the dirt, discoloration and spalling caused by Victorian urban pollution. Advocates described it as *"practically time-proof and indestructible."* In their RIBA lectures, Sir Charles Barry and others advocated terracotta's "self-cleaning" property in contrast to stone, which attracted soot and grime. In 1868, Gilbert R. Redgrave proclaimed: *"in good terra cotta we have a material which defies our destructive climate."*[14]

In this same period, Lewis Gruner and his team documented the fourteenth-century churches of Pavia and Crema in Lombardy and remarked on the crisp appearance of the terra cotta.

In the United States, the noted architect James Renwick commissioned terra cotta façade elements from a sewer pipe factory and substituted them for cut stone on a series of New York projects, beginning in 1853. Seventy-two years later a retrospective on the work was published in *The American Architect*:

> Every stonecutter and mason in New York told Mr. Renwick that terra cotta would not last – in spite of the evidence in northern Italy. To realize how badly mistaken they were it is only necessary to compare the condition of the terra cotta he made with contemporary brownstone

[11] Stanford, p. 107.
[12] Stanford, p. 113.
[13] Henry et al., p. 674.
[14] Henry et al., p. 675.

... the terra cotta is unchanged. The modeling is as fresh and crisp, though somewhat Victorian in detail, the lines as clean-cut as on the day the terra cotta came from the kiln.[15]

The resistance of terra cotta to moisture-related weathering effects is due to the relatively low porosity of the material. The historical range is considered to span from 5% to 10%. Porosity is determined by the degree to which the clay vitrifies in the kiln. The fine particles that gather at the surface of the formed clay will vitrify to produce a "fireskin." This skin has additional protective benefit, however it is not impervious. Vitrified, glazed surfaces can reduce porosity to less than 1%, although it will increase with long-term exposure.[16]

Moisture becomes a greater threat to terra cotta if it freezes, expanding in volume within the pores of the material. Repeated freeze/thaw cycles induce tensile stresses in the material, leading to internal weakness and possible spalling of the surfaces. Building tops and projecting elements exposed to more water and temperature variation are more challenged. However, the specific response of terra cotta is a multi-variable phenomenon.

> The susceptibility to frost failure depends on the clay type and manufacturing process: terracotta units with very different characteristics may be equally resistant, though for different reasons. For example, a dense block with a high compressive strength will be resistant because it absorbs little water; whereas a weak, highly absorptive block can still be resistant, because in this case the open pore system both dries quickly and allows space for the ice to expand. Low-strength or under-fired terra cotta tends to be the most susceptible.[17]

Beyond moisture related effects, terra cotta is known for superior resistance to fire. The historical expectation has been that clay products are burned in the kiln at temperatures exceeding those generated by common fires, and therefore combustible impurities within them have already been consumed. The ability of clay components to control fire spread would have been easily observed wherever they were used. A culture of brick building was developed or enforced in cities that suffered historic fires, from Nero's Rome (64 CE), to London (1666), to the village of Nantucket (1846) off the Massachusetts coast.

In truth, the firing temperature of clay products varies with the types of clay. Kaolin and ball clays fire at relatively low temperatures (800–1160°C) (1472–2120°F) and are valued for their plasticity and their white or pale body colors. Fireclay and marls are not nearly so plastic, but they shrink far less and can be fired at much higher temperatures (1170–1350°C) (2138–2462°F). They produce classic, red body colors that range darker according to the iron content, and they offer greater fire resistance.[18]

[15] *The American Architect*, November 20, 1925, p. 429, quoted in Tunick, p. 6.
[16] Henry et al., pp. 656–7.
[17] Henry et al., p. 700.
[18] Henry et al., p. 640.

Terra cotta construction techniques evolved in the consciousness of fire risk. First uses as roofing in ancient cities protected combustible buildings from falling embers. Tile roofs spread across Europe in the middle ages and by 1212 they were required in the City of London.[19] The London Building Act of 1774 outlawed the use of wood for porches and ornament. Although the immediate response was the use of stone, the fire regulations created market opportunities for less expensive Coade Stone.[20]

The Chicago fire of 1871 is often cited as the defining catalyst that stimulated expansion of the American terra cotta industry. This is not to say that fireproofing techniques using clay were unknown before, nor were they universally adopted after. A second conflagration, in 1874, inspired Chicago's first comprehensive fire code.[21] Nevertheless, the more famous Chicago fire is used as the benchmark.

The Cooper Union Building in New York City was the first documented use of hollow clay tile in the United States. It had been built in 1853 by Frederick A. Peterson, but it was considered an anomaly.[22] Before the Chicago fire came public warnings from architect P.B. Wight about the vulnerability of unprotected iron structural members that had begun to proliferate.[23] Massive evidence of burst stone and twisted iron beams in Chicago pushed Wight's concerns to the forefront. Hollow clay floor systems quickly evolved as part of the solution. Fitted between iron beams, tiles held a significant advantage over brick arches because of the reduction in weight.

The Chicago Terra Cotta Company had just opened in 1869. The Nixon Building in Chicago survived the great fire, and the public notice that it drew stimulated the growth of iron structural systems fireproofed with clay. The pioneers of fireproofing in New York extended their business interests to Chicago.[24]

Numerous factors contributed to the creative explosion that took place in Chicago in the decades that followed. It would be difficult to overstate the role terra cotta played in the development of "high rise" structures with a uniquely American character. Thomas Leslie's book, *Chicago Skyscrapers: 1871–1934*, offers a full account. It is organized around the emerging technologies that drove the development rather than the, often confusing, chronology of individual events.

These [Chicago] entrepreneurs made improvements in four areas: understanding the distinction between *incombustible* and *fire-resistant* construction; finding ways to contain and compartmentalize fires inside and out; fireproofing metal columns in addition to beams; and improving existing terra cotta floor systems. The distinction between *incombustible* and *fire-resistant structures* reflected a growing awareness that

[19] Davey, p. 159.
[20] Henry et al., p. 60.
[21] Leslie, p. 11.
[22] Wells, p. 28.
[23] Tunick, p. 8.
[24] Leslie, p. 16.

an iron or timber structure itself was only part of the problem. Wooden trim, wall studs, flooring, doors and furnishings were all part of any office building, and they all offered fuel for a developing fire.[25]

Inventors George Johnson, Balthasar Kreischer and William Freeborn, among others, improved the earlier approaches to fireproof floor spans. Ultimately, they wrapped iron beams in a compact assembly of extruded terra cotta shapes that protected them from the heat of a fire. In addition, these systems produced flat surfaces above and below that facilitated the completion of floor and ceiling finishes without layers of furring that would contribute to the fuel load.[26] Hollow clay partitions were used to contain potential fire sites, finished with ceramic materials to further reduce the amount of wood. Of particular interest, from a technical perspective, was the development of porous, nailable, terra cotta "lumber" for use as incombustible trim.[27] In his practical treatise of 1884, Charles Thomas Davis described the production of this material:

> The New York Terra-cotta Lumber Company has established large works at Perth Amboy for the manufacture of lumber by mixing resinous sawdust with the wet clay, which is left porous after burning by the sawdust being consumed. … It can be applied to a variety of uses; it is light, bulk for bulk, and may be united like joiners' work or nailed into place like so much wood.[28]

The building systems in Chicago continued to evolve, with an expanding role for terra cotta, valued not just for fire resistance but for its relatively light weight. On the exterior, terra cotta facilitated a new architectural language for curtain wall façades that were fully supported by the frame structure behind them. The steps in this transformation are elegantly presented in Thomas Leslie's account. Burnham and Root's Reliance Building of 1895 is the benchmark of a most formative stage. It was followed by the Fisher Building, of 1896, in which the principal means of construction were steel, glass and terra cotta, with very few bricks.[29]

The 1906 San Francisco earthquake and fire destroyed numerous iron and steel buildings. They had been marketed as fire protected, but later code revisions increased the terra cotta covering from a mere 1.5 inches (38 mm) to 4 inches (100 mm) in thickness.[30] As rebuilding progressed, concrete took over the construction of columns and floor spans, combined with terra cotta for infill and cladding. The place of the material was secure in the recovery and growth of the city because it had by then become accepted by architects and appreciated by the public for its versatility as a finish surface in foyers and façades (Figure 2.1).[31]

[25] Leslie, pp. 16–17.
[26] Leslie, p. 16.
[27] Leslie, p. 17.
[28] C.T. Davis, pp. 308–9.
[29] Leslie, p. 96.
[30] Kurutz, p. 95.
[31] Kurutz, p. 95.

Figure 2.1 Elevator lobby at 450 Sutter Street, San Francisco (1929). Designed by Timothy Pfleuger. Glazed terra cotta by Gladding McBean Company, Lincoln, California. Photo: Donald Corner.

The most poignant story of buildings from this era is that of 90 West Street, located at Ground Zero in New York. Cass Gilbert's landmark structure from 1907 is considered a "miracle building" because it survived the collapse of the adjacent World Trade Center towers, while other buildings did not. Debris rained down on the structure, gouging out sections of the façade, and setting internal fires that persisted for five days. The 23-story prototype "skyscraper" had 4 inch terra cotta blocks protecting the columns, and 12 inch (30.5 cm) thick terra cotta floors.[32] The interleaved blocks on the elevations survived the heat of the fires, excepting the areas of direct hit. Boston Valley Terra Cotta reproduced the damaged parts, including an extra copy of a gargoyle for the upper floors (Figure 2.2). That creature stands watch over their shipping room as testament to the product and in memory of the human tragedy of 9/11.

Contemporary terra cotta responds to a complex network of protocols for quality assessment. First, it is traded internationally, so specifications and testing methods at the point of manufacture may differ from those at the point of end use. Second, there are distinct domains within the overall spectrum of clay products, and each must meet different criteria. Traditional, hand pressed "architectural terra cotta" does not face the same challenges as extruded "rainscreen cladding." Third, in terms of volume, terra cotta is a relatively small segment of the clay products market. It operates under protocols designed for the large volume items. In the United States, testing protocols first flow from ASTM C67/C67M-20: Standard Test Methods for Sampling and Testing Brick and Structural Clay Tile. This includes protocols for: compressive strength, modulus of rupture, absorption, saturation coefficient, freeze/thaw resistance, and variables related to weight, size, length change and void areas.[33] Several other ASTM standards can be used to analyze flexural properties (modulus of rupture and modulus of elasticity), including some originally designed for dimension stone.

Architects can work within systems and techniques that have performance histories established by the producer. If they choose to stretch those limits, they should rely on consulting engineers with a particular expertise in terra cotta to chart a path through the applicable standards. Some tests are readily performed and can be project specific. Others are of long duration and depend on the production history of a chosen manufacturer. An example of the first is the testing of flexural properties for rainscreen panels that span from one mounting clip to another. In certain jurisdictions they must resist hard body and soft body impact tests, as well as wind load.

[32] Collins.
[33] ASTM International.

Figure 2.2 Ornament on the upper floors at 90 West Street, New York, NY (1907). Architectural design by Cass Gilbert. Damaged by flying debris from the World Trade Center collapse, the façade was repaired with new elements by Boston Valley Terra Cotta. Photo: © Christopher Payne/Esto.

In the latter category is testing for freeze/thaw effects. In the major market of New York, codes require terra cotta samples to survive 300 freeze/thaw cycles, the equivalent of thirty years in use. At the lab, it takes twenty-four hours to complete each cycle, so this is not a test that can be commissioned anew for every project. It depends instead on the integrity and stability of the manufacturer with respect to clay composition and processing. The model specification of the New York City School Construction Authority is recommended to architects by producers.[34] It permits waiver of a specific, production run test if there is certified data for the same clay and glazes that is no more than two years old.[35]

The architect's role in the design of a cladding system need not involve full engineering, exhaustive detailing and highly specific testing. In many cases the architect will describe the design intent, including sizes, profiles, finishes and dimensional requirements, and identify a "basis of design" terra cotta product. Typical testing parameters identified above provide quality expectations to inform the selection of a qualified installer and manufacturer who further develop the system. This is known as "delegated design" or "design build" using the manufacturer's engineering and detail standards.

Experts in the North American market have discussed the need for standards that are specific to terra cotta cladding, but they have not yet been formulated. In the case of freeze/thaw, the goal would be a predictive test that reveals how long terra cotta can be expected to serve.[36] Proving it will survive thirty years of simulated exposure undervalues the historic durability of the material. Designing buildings for longer service is central to a sustainable future.

[34] Farnham.
[35] New York City School Construction Authority, section 04250.
[36] Farnham.

CHAPTER 3

Form and Ornament

Architectural terra cotta presents the opportunity to seamlessly integrate performance-driven form with aesthetic content developed as ornament. All primary materials – wood, stone or metals – do this to some extent, but none more readily than burned clay. From pre-history it has been a material of pragmatic, domestic utensils and a material of refined sculpture. It can be readily worked by hand while wet, easily carved before drying, and milled to a final form after firing. In the words of the French archaeologist and historian Jean Baptiste D'Agincourt: "Soft clay being of all substances that which lies most ready to the hand of man, could not but be found most available for an infinite number of objects of first necessity."[1] First necessity might include cooking pots, or vessels for containing liquids. It might also refer to a sheltering roof, with curved clay tiles as one of the fundamental building components made from this material. Techniques for working clay begin with a legend. Slabs are said to have been formed over the thigh of the worker, producing both the curve and the taper required for a lapped roof.

Pragmatic objects were enriched by artistic transformation of the basic components. The pouring spout of a pot or the handle on a lid could become representations of domestic animals or mythical creatures. Architectural elements responded to the same impulse. Etruscan remains on display in Murlo, south of Siena, include ridge tiles augmented at each bay with an acroterial statue known as the Murlo Cowboy (Figure 3.1). Similarly, the antefix is a sculptural addition to the eaves of a building, an end to the cover tile at the base of the slope (Figure 3.2, top image).

The figural tradition is evident in remnants of the Temple of Mater Matuta (sixth century BCE), on display at the Capitoline Museums of Rome. Clay revetment plaques protected the timber structure of the roof as well as providing a narrative in the ornament (Figure 3.2, bottom image). Clay art

[1] *Recueil de Fragments de Sculpture Antique en Terre Cuite*, Paris, 1814, as quoted in Gruner, p. B.

DOI: 10.4324/9780429057915-4

Figure 3.1 Acroterial statue from the ridge of the archaic building at Poggio Civitate, an Etruscan site located at Murlo, (Siena) Italy. The structure was rebuilt circa. 580 BCE. The figure is known as the Murlo Cowboy. Photo: Donald Corner.

Figure 3.2 Above: Palmetto antefix (50 BCE–50 CE). Displayed at the Palazzo dei Consoli, Gubbio (Perugia) Italy. Below: Revetment plaques from the Tempio di Mater Matuta, sixth century BCE. Displayed at the Capitoline Museums, Rome, Italy. Photos: Donald Corner.

reached a level of mastery in the work of Lucca della Robbia in the middle of the 1400s. A significant advancement was the brilliant, durable glaze that is credited to the artist. Large applications, such as the lunette of a portal, were composed of close fitting of parts small enough for the kilns, with meeting lines that worked around the bold figures to minimize their impact.

METHODS OF FORMING

The first and most basic method of forming clay is hand pressing. The fundamentals carry forward from the era of the Della Robbias to the classic period of architectural terra cotta. In the United States that period stretched from the 1880s through the 1920s. Production in this era is chronicled in the writings of Walter Geer[2] and Susan Tunick.[3] The remaining American firms are actively engaged in hand pressing replacement parts for historic structures. As the craft is preserved and advanced, it is also available for use in contemporary work.

Still visible at Gladding McBean are tilt-tables that allowed artisans to work on large clay masses that would become sculptural panels or bas-reliefs over a significant expanse of the building façade. Like all clay components, the figures were designed and modeled at a larger size to account for shrinkage while drying and firing. The originals were cut into sections that were manageable through the rest of the process. The limits of size at a given facility were influenced by what else they produced: bricks, roof tiles or clay pipe. Gladding McBean still produces enormous clay pipe sections. Their architectural portfolio at peak production reflected this same scale and capability.

Written accounts describe how pieces of solid originals could be hollowed out at the back, then dried and fired directly.[4] Boston Valley fired original sculptures when replacing the gargoyles on Harris Hall, at CCNY in New York.[5] Under most circumstances, the investment in an artist's original modeling is protected by making a plaster cast of each piece. These molds are then hand packed with carefully prepared clay of uniform density, free of air bubbles. This offers security, even when only one successfully fired unit is required.

Gargoyles or finials at the top of a building may be more resistant to extreme exposure if the critical surfaces are free of joints. The clay sculptor produces what is known as the "model" or "first positive." Rather than cutting this up with a taut wire and making simplified molds of each piece, the choice can be made to build the mold in multiple segments that surround a single large positive (Figure 3.3). If the original form is undercut, the mold has to be logically subdivided, allowing it to be sequentially removed from around the green clay.

[2] Geer.
[3] Tunick.
[4] Elliot, p. 54.
[5] Fritz.

Figure 3.3 Construction of a segmented plaster mold around a sculpted figure at Gladding McBean Company, Lincoln, California. Photo: Donald Corner.

The extraordinary reach of these techniques is demonstrated by the recent work at Boston Valley (2017) to replicate the ornamental cartouches designed by Louis Sullivan for the Gage Building in Chicago.[6] The recovered original components include flat backed, ashlar units with boldly cantilevered plant forms clinging to the face. The interlocking parts were duplicated in the model shop and trial fitted to check the alignment of tendrils that curl from one piece to another. In the mold shop, infills are added behind the deeply undercut elements to support them and rationalize the ultimate shape of the surrounding plaster. After the multi-part molds are removed, the supporting infills are cut away again. The entire assembly is trial fitted both before and after firing to confirm that the parts will go together in sequence (Figure 3.4).

The formal language of architectural terra cotta owes a great deal to its historical use as an economical replacement for carved stone string courses, window surrounds and cornices. The predominance of rectilinear volumes derives from blocks of stone, and this greatly facilitates their reproduction in clay. Working with the first positive, a simple mold box is constructed with the inverse of the primary face at the bottom. Four detachable sides are used to develop the depth of the unit

Figure 3.4 Ornamental elements of Louis Sullivan's façade at the Gage Building, Chicago, Illinois (1890–9). Replacement parts (above) compared to original components recovered from the building (below). Photo: Boston Valley Terra Cotta, Orchard Park, New York.

[6] Fritz.

(Figure 3.5, top image). All five of these surfaces are packed with clay to a specified thickness. Webs are formed across the void to strengthen and stabilize the shape through the rest of the process. Warm air is blown into the cells from above to begin the drying process, allowing the clay to shrink away from the mold face so that it can be disassembled (Figure 3.5, bottom image). Turned over on to a palette, the piece is sent on for hand finishing of the visible surfaces before it moves to the drying shed (Figure 3.6).

The more regular the form, the more strategies available to generate the first positive. A continuous profile used to be "run" in plaster by drawing a sheet metal template down the length of the bench, progressively shaping the wet mix (Figure 3.7, top image). This approach points the way to the discipline of extruded forms to come later. By whatever means the first positive is started, the modeler can add ornamental relief within the margins, usually with plasticine rather than clay. The combined result can then be replicated by hand packing clay into plaster molds made from this original (Figure 3.7, bottom image).

At Boston Valley, digital media have been integrated into this process. Photogrammetry is used to build a 3D model of each historical part, whether it is available in the shop or remains on the building. From this model a simple substrate can be defined. Blocks of foam are cut to sufficient size with a wire cutter and then passed on to a CNC router for refinement of the profile. Fine detail is added by the modeler to complete the first positive (Figure 3.8).

Clay for hand pressing is prepared in slabs that are literally punched against the form face in order to fill all of the sculptural detail and eliminate air pockets. When de-molded, the surfaces are not perfect. They must be filled and smoothed, which is the greatest single labor component in the process. The details must be cleaned and straightened if they have been distorted coming out of the mold. Fine grain, tooled textures that are too subtle for the mold are added by hand. It is also possible to extend a few undercut projections that defy direct molding (Figure 3.9, top image). This work is done with the element face up, supported by a consistent, symmetrical pattern of webs and voids at the back. The final piece may not be symmetrical if it wraps around a support armature. The extra clay is outlined with a kerf and marked for removal after drying and firing (Figure 3.9, bottom image).

After the pieces are quality controlled and stamped for identification, they are set aside on carts or palettes to begin drying. When they reach a state referred to as leather hard, they move on to the drying sheds. It is

Figure 3.5 Above: A five-part plaster mold for hand pressed clay. Below: Preliminary drying of the pressed clay using hot air hoses. Gladding McBean Company, Lincoln, California. Photos: Donald Corner.

Figure 3.6 Finishing of a de-molded, hand pressed clay component. Gladding McBean Company, Lincoln, California. Photo: Donald Corner.

Figure 3.7 Above: Plaster horse used to "run" the body of a linear architectural element. Below: Modeler adds detail to the "first positive" from which molds will be made. Gladding McBean Company, Lincoln, California. Photos: Donald Corner.

Figure 3.8 Making a new model from a recovered historic part, scaled up to allow for clay shrinkage. Modeler adds ornament to a machine cut foam block. Boston Valley Terra Cotta, Orchard Park, New York. Photo: Donald Corner.

critical that each piece dry completely, from the inside out. If the surface dries first and water is trapped behind, the piece can explode in the kiln. To control the drying process, hand pressed units have a uniform wall thickness (1-1/4 inches or 32mm) regardless of their overall size. The drying shed operates at 100°F (37.8°C) and 90% humidity to prevent rapid surface drying. Pieces stay in the shed for five days to five weeks depending on their size and configuration.[7]

Hand pressed units vary tremendously in size and shape. They are usually fired in an intermittent box kiln or shuttle kiln to accommodate this range of dimensions. For efficient use of the kiln, the pieces are organized on a cart or table with multiple layers supported by material that can withstand the firing temperature. Generally, the carts move into the kilns, but at Boston Valley the kiln itself moves over the loaded table so that the rack of unfired pieces is not shaken in transit. Once the kiln is closed over the ware, it remains in place for the full heating and cooling cycle before it is opened up and moved on to another batch.

The second, basic means to form clay is by extrusion; pushing a column of clay through a steel die to produce virtually any imaginable, continuous profile. Combining design versatility with production efficiency, extrusion is

[7] Boston Valley, 2019.

Figure 3.9 Above: Projecting element added to a hand pressed clay block after de-molding. Below: Kerfs and markings indicate portions of a block to be removed after firing. Boston Valley Terra Cotta, Orchard Park, New York. Photos: Donald Corner.

the predominant contemporary method. Of added significance is the ability of clay to flow around an array of cores suspended in its path, and to fuse as a continuous body on the other side. This provides for substantially hollow cross sections with a network of web members that strengthen the whole (Figure 3.10, top image). Closed and cross braced forms are particularly important in clay because, unlike metals, the extruded elements have relatively limited strength until much later in the process. All extrusion processes favor a uniform design thickness for the material within the section so that the rate of flow through the die will be consistent and therefore free of internal stresses. This principle has an additional benefit in clay because the consistently slender side walls of the cellular form are easier to dry and fire in the stages ahead. Finally, the pressure of the extrusion produces a clay body of uniform high density, and this translates into strength and durability in the final product.

Clay extrusion was developed for prosaic products before it was adopted for architectural terra cotta. A steam press that extruded finite length of clay pipe was patented in England in 1848.[8] In 1863, Cyrus Chambers of Philadelphia patented a brick machine that used an auger to extrude a continuous column of clay.[9] In the United States, structural clay tiles were formed with steam presses before 1900 and shifted to auger feeds after that.[10] Architectural terra cotta in the early twentieth century was often made in facilities that featured volume production in these other sectors. Foremost among them was structural clay tile, with complex cross sections that pointed toward future possibilities in architectural work.

The extrusion of architectural terra cotta finally gained popularity in the 1920s.[11] There are two widely recognized catalysts for this change: the desire to reduce the labor requirement of hand pressing in a postwar economy, and a movement in architectural design toward low relief (Art Deco) ornamental motifs, in contrast to the voluminous and deeply undercut elements of preceding styles. In her account of downtown Portland, Oregon, *Last of the Handmade Buildings*, Virginia Guest Ferriday evokes a sense of loss in the move away from hand pressed terra cotta. The range of form and expression was unmatched, and there is an undeniable pejorative in the term "machine made terra cotta" applied to everything that followed. Nevertheless, it was the shift toward extrusion that laid the groundwork for subsequent revivals of the material.

Design intent for extruded clay products is encoded in the construction of the dies. Bricks have a dimensionally coordinated form, so relatively few, simple dies can deliver a huge number of units. Special bricks at the corners may require a different die, fitted with moveable parts so that one die can produce more than one corner solution. The fewer the dies, the lower the unit cost of the delivered product, due to the reduction of front-end investment.

[8] World's Oldest Industry, p. 21.
[9] Chambers, sheet 1.
[10] Wells, p. 32.
[11] Tunick, p. 30.

Figure 3.10 Above: Extrusion of a terra cotta rainscreen plank. Below: Storage of extrusion dies for hollow core planks of various profiles. Palagio Engineering, Greve in Chianti (Florence) Italy. Photos: Donald Corner.

Contemporary terra cotta façade planks can be made with a similar approach. If a limited number of plank profiles are offered, there can be a library of prepared dies ready for use (Figure 3.10, bottom image). The face dimensions, edge shapes, core pattern and wall thickness may be used to define each family of products. The leading producers refined this approach to wall claddings over a period of several decades, beginning in the mid 1980s.

Today, there is so much competition in the production of standardized shapes that experienced firms have turned away from a high volume approach, toward high quality, custom designs for individual clients and projects. A single, high profile, international project completed in 2019 by Palagio Engineering used 43 different die shapes to make 650 distinctively different tiles.[12] This degree of customization requires sophistication in the design of the die holder, creating a simple interface between the general conditions and the specific parts that must be changed to create a given profile. The die itself is a steel plate cut very precisely to describe the outer boundary of the desired section. There are unique plate elements that must be held in the middle of the clay column to carve out the inner cores (Figure 3.11). The cutting of dies is done with digitally controlled devices that directly apply design files approved as shop drawings. Machining of

[12] Palagio Engineering, 2019.

Figure 3.11 Die inserts for custom extrusion profiles with hollow cores. Palagio Engineering, Greve in Chianti (Florence) Italy. Photo: Donald Corner.

tool grade steel has long been automated to some extent, but the digital environment increases this capacity, while controlling the costs.

The maximum sizes of an extruded clay column vary from one producer to another. With this variation in the volume of clay being handled and the pressures applied, there is a corresponding variation in the weight, strength and cost of the die components (Figure 3.12). Value results when there is a good match between the sizes of the units desired and the operating range of the preferred providers.

The first order of inquiry is to determine the efficient heights and widths that can be delivered by the extruder. This range can be described as a rectangular portal through which the designed profile must pass. The aspect ratio of the rectangle can vary, from broad, flat sections to those that are more nearly square. For the most common extruded shapes, one side will be considered the unseen back, as it is the surface that will come in contact with rollers that convey the pieces through the rest of the process. All the other faces can be of the highest quality since they are not marred by contact once they emerge from the die. In some facilities it is possible to rotate the die holder 90 degrees so that a short side is the "back" rather than the more common broad side. It is also possible to specify shapes that are true from the die and unblemished on all four faces. These are referred to as tubes, louvers or baguettes, and can include surprisingly large sections. The techniques used to produce them are often proprietary. Specific examples will be presented in Chapter 6.

The overall dimensional limits are no more than a place to start. The final cross section cannot simply fill the whole rectangular window. There are many other variables to work through: the length of the piece, the weight that will result, and the allowable tolerances for distortions that will occur within irregular shapes. There may be control devices inserted behind the die to regulate the flow of the clay across different parts of the section. As the clay emerges from the extruder, it feeds onto a table of rollers set at the correct height to receive the bottom, or back of the section. Pieces requiring special handling may be fed on to sheet metal trays. The column is cut into lengths with a wire while it is still in motion. At some facilities the individual pieces may be cut again, while stopped, for greater accuracy. A final cut will be made after firing.

The permissible length of a section also varies by producer, influenced by the control they can maintain over the warping of a piece. In the early days of contemporary façade planks these lengths were quite limited. With improved process technologies, flat planks can now be much longer. The

Figure 3.12 Large-scale extrusion die with cross bridges to hold internal core profiles. Above: View from downstream, clay exit. Below: View from upstream, clay entrance. Gladding McBean Company, Lincoln, California. Photos: Donald Corner.

limiting condition may not be distortion, but rather the width that has been chosen for the drying carts or the kiln.

In a modern facility, green clay segments are fed automatically onto the various levels of a rack that next moves into the drying chamber. Fans move over the height and length of the enclosure, blowing air through the ends of the extruded shapes. The ware is dried relatively slowly over a period of two or three days, reaching a temperature of 70–100°C (158–212°F). In some cases, water is sprayed on the exterior surfaces of a complex section so that the rate of drying is uniform from both inside and out.

After passing through a spray booth for the application of glaze, the ware moves along to a linear kiln. Contemporary kilns are organized in three stages: a zone that progressively pre-heats the work, a zone in which the designed firing temperature will be reached, and a zone for progressive cooling. In a roller kiln the pieces are borne along on closely spaced, rotating tubes of an industrial ceramic capable of withstanding the firing temperatures. Such a kiln may be 120 meters long. There is only one layer of rollers, and therefore one layer of product moving under the heat source. Characteristic firing reaches approximately 1100°C (2012°F). The duration of the journey through the kilns depends on the size and shape of the pieces, with the shortest passages being under 15 hours.[13]

Tunnel kilns, common in brick production, require that multiple layers of product be stacked on a heavy cart that has a fire brick bed to prevent melting of the steel wheels and rails below. Terra cotta pieces need not touch each other in the vertical stack as a matrix of fire bricks and the same industrial ceramic tubes can be used to support each layer. Such a kiln may be 70 meters long and, moving slowly, the passage may be completed in three days. However, multiple layers of product are moving through at the same time.

Slip casting of clay offers extraordinary design freedom in three dimensions, without the linearity imposed by the extrusion process. It is a simple technology with very little capital investment required. This necessarily implies a low output rate in architectural applications, suitable for featured elements rather than the broad expanse of a façade.

The word "slip" in this usage refers to a liquefied suspension of clay particles in water, generally thinner than a slurry, and formulated for this purpose as a "casting slip." When poured into a gypsum plaster mold, the slip interacts with the mold face, exploiting the ability of gypsum to draw water out of the mix. The clay dries and solidifies, beginning at the mold face and progressing inward over time. The process can be allowed to continue until the internal volume of the mold is filled with solid clay.

[13] Lehmann, 2020.

In classic applications, the process is stopped after the designed side wall thickness has been reached. The remaining liquid is drained out of the center through a hole, leaving a hollow slip cast (Figure 3.13).

The first positive for slip casting is a solid object sufficiently concentric in form that a mold can be constructed around it free of undercuts or interlocking parts that would prevent de-molding. The positive can be produced by any conceivable method: carving, sawing, turning on a lathe, or CNC routing directly from a digital design file. The original is embedded in gypsum plaster, artfully divided into parts that can be sequentially removed to free the interior object. The mold joints lock together to contain the fluid material and produce an accurate replication of the form. From the exterior, the mold presents as a cubic block of gypsum, bound together by straps, with holes in the top to fill and possibly empty the slip.

A slip cast mold may be opened in as few as eight hours, then prepared for re-use. After ten uses, the gypsum faces become saturated, and the

Figure 3.13 Left: Gypsum plaster mold, with top removed, prepared for slip casting. Right: De-molded ridge crest segment with overlapping joint cover at far edge. Prepared for The Old Red Courthouse, Dallas, Texas at Boston Valley Terra Cotta, Orchard Park, New York. Photos: Donald Corner.

mold must be dried out before it can be used again. If a large number of unit copies are required, multiple molds can be cycled concurrently so that the whole project moves through the plant in a compact time frame.

Hollow castings deliver potentially large, durable building components that are light in weight. A rooftop balustrade made of hollow clay units can easily outperform cut stone, or cast stone (precast concrete), which are much heavier. However, the hollow vessel must have sufficient folding, curvature or thickness to retain its shape in the kiln.

Cladding the walls of a building usually allows even complex shapes to have a distinct back. The piece can be reduced in weight using a network of coffers and ribs on the back side without internal hollows. In this case, the clay in the mold is allowed to dry completely to form a solid slip casting. Liquid clay can be introduced into a mold with less effort than hand pressing. It can flow around corners to create dramatic, three-dimensional objects, provided there is a way to sequentially remove the form. Since the clay fills the mold, the thickness is precisely controlled by the mold design and all of the surfaces are true to the dimensions of the original positive. Occasionally a piece is hollowed out on the back or bottom to the extent that it cannot support itself in the kiln. Then, it must be held up by a refractory material to prevent slumping or warping. The logical choice for a custom support is to also make it of fired clay, perhaps also slip cast.

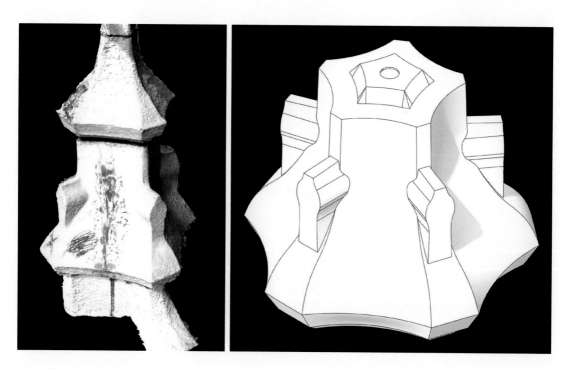

Figure 3.14 Development of replacement parts for the Woolworth Building, New York, NY. Left: Output from scanning process. Right: Digital model of the intended part. Images provided by Boston Valley Terra Cotta, Orchard Park, New York.

The technical simplicity of the molding process makes slip casting the logical choice for special conditions: corner, edges, openings and ornament. The production of "in-kind" replacement parts demonstrates the versatility of the method. For Boston Valley Terra Cotta, damaged parts can be scanned in place or removed for scanning at the facility. The captured data is used to produce a 3D digital model (Figure 3.14), which can be converted to a physical model using CNC equipment to shape a foam block. This first positive is reproduced in clay using slip casting when the parts required are complex in form but limited in number (Figure 3.15).

Ram pressing wet clay produces outcomes similar to those of solid slip casting, but with a more industrial approach. A block or slab of plastic clay is compressed between two die faces using a hydraulic ram. The cycle time is much shorter for a press, lending to greater volumes of production. In return, there are limitations on the forms that can be created as the upper, female die is lowered over the male die and then raised again along a single axis of movement. Where multi-part molds can be disassembled to release complex forms in slip casting, they are contradictory to the notion of rapid cycling in a press. The vertical dimension of a pressed piece is limited by

Figure 3.15 Left: Sample part recovered from the Woolworth Building, New York, NY. Right: Completed preplacement part. Photos: Boston Valley Terra Cotta, Orchard Park, New York.

the draw depth of the ram and the working limits of the material. There is a draft angle imposed on deeply folded forms. Looking at the projecting elements of the male die face, the sides will taper by 2–4 degrees. This allows the reciprocal female die to move away from the formed clay as it is raised, rather than shearing it between two faces that are parallel to the axis of the ram.[14]

The ram applies high pressure to the clay, relative to hand work, but relatively low pressure compared to other industrial materials such as sheet metal. This means that the molds, or dies, do not need to be extraordinarily strong. Gypsum plaster is the logical option for the die face. To create a die, plaster is cast over solid block-outs resting on a base plane. The plaster is contained at the sides by a re-usable steel box with exterior fittings to attach it to the press. That much is easy. However, the clay will be pressed against the face of the gypsum and if the machine is to cycle rapidly there is no time to allow it to shrink back from the surface to break the bond. Instead, the steel box must be fitted with a network of tubing that delivers a water borne release agent to the die face. The distribution tubes are embedded in the cast plaster and rely on the porosity of the gypsum itself to complete the delivery process. The box may also be spanned with lengths of rebar welded to the boundary wall and used to hold the shell and gypsum together. All of these embedded elements make the ram die expensive to fabricate. By rule of thumb, each one must produce fifty finished tiles to justify the use of this method.[15]

When the female die is lifted by the ram, a pallet is slid underneath and the suspended clay part is released on to the pallet for further handling. The pallet may include shims that support overhanging clay elements, like shiplap joints, that would otherwise slump before drying. The exposed face of the unit is trimmed and touched up by hand. This requires more manual intervention than an extruded part, but far less than hand pressed clay (Figure 3.16).

POST PROCESSING

Clay products that have emerged from a forming process can be easily altered or augmented while the material is still in a plastic state. The simplest types of alteration include the filling and smoothing of voids left on the surface of a hand pressed unit or trimming off excess clay at the meeting line between two mold parts (hand pressed) or two die faces (ram pressed). More complex operations include the refinement of sculptural detail as described in the hand pressed section above. The wet clay offers a perfect opportunity to add texture, and those possibilities will be addressed in Chapter 4.

[14] Gulling, p. 336.
[15] Boston Valley, 2019.

Figure 3.16 Ram pressed panels for the Asian Art Museum of San Francisco, California. Above: Released from the mold onto pallets, headed for drying. Below: Detail of support for shiplap edge while in green state. Boston Valley Terra Cotta, Orchard Park, New York. Photos: Donald Corner.

Bolder examples of post processing substantially change the shape and volume of the original moldings. Two independently formed pieces can be added to each other and move forward as a single, structurally integrated unit. This is the timeless art of adding a handle to a coffee mug. In volume production, the elements to be combined or altered through manual intervention will begin with systematically formed base components, as far as possible. For example, to develop a right angle on a section of shiplap plank, two lengths of green clay can be miter cut and then rejoined at the corner. Hand smoothing will be required on the visible, outer surface.

To produce a tight radius corner that is compatible with the wall system on either side, a section of extruded plank can be hand pressed into a gypsum plaster mold (see Figure 6.57). Accurate replication of the curve is assured by the mold and the cross section at the ends will match the adjacent, standard panels. This new, three-dimensional element will no longer lie flat with the exterior face up as it passes through the kiln. If left unsupported, the clay will slump unpredictably. One approach is to add clay web members on the back to provide a stable base, cutting them away after firing. More efficient is the use of refractory supports; a set of previously fired clay sections designed to retain the final shape.

The tendency of clay to slump while in a plastic state can be used to create large radius, curved shapes (Figure 3.17). Each producer has a different approach to this process, and for most the sequence of operations and exact techniques are considered proprietary. An extruded clay plank can be slumped over a large radius form before it is thoroughly dried. To maintain a consistent outcome, the designed curvature must again be reliably supported during the firing process, as with a refractory arch. Large pieces and their supports may be loaded in tiers for passage into a shuttle kiln. Alternatively, special support configurations can be created within the racks that are used to stack pieces on the carts of a tunnel kiln. Through the design of the die, and control of the flow rates through the extruder, the clay column can be prepared so that it "wants" to distort into the desired curvature, and will do so without excessive stress. After firing, the curved sections are supported with a wood form while they pass through an automated sawing station to produce the final lengths and angles at the end cuts. Hand interventions can be used to make dramatic, three-dimensional shapes such as the twisted panels on the façades of the Voxman School of Music at the University of Iowa (2016) (Figure 3.18). An extruded length was gently pressed against a custom back board that was developed from digital models with the assistance of the architects.

Figure 3.17 Rounded corners formed from extruded sections. Above: Sprayed with glaze and ready for a second firing. Below: Completed units shown with quality control template to verify the final shape. Palagio Engineering, Greve in Chianti (Florence) Italy. Photos: Donald Corner.

Figure 3.18 Voxman Music Building, University of Iowa, Iowa City (2016). LMN Architects, Seattle, Washington. Four sided, extruded panels, 1–1/2″ x 8″ [38 x 200 mm] in section, are slump formed with a 90° twist over their full height. Photo: Tim Griffith.

Extruded sections are usually closed forms to provide stability and re-silience during the drying and firing processes. These shapes can be cut apart to create one or more finished surfaces for application as a veneer. The classic version of this technique was the so-called "machine made terra cotta" of the 1920s and 1930s. Large-scale extrusions were used to make substantial rectangular sections. The cross webs helped to maintain true, flat faces that would otherwise have warped while firing. Cutting the section in half, after firing, and removing most of the webs produces two completed elements that are larger in area, yet thinner and lighter than would otherwise have been possible at the time. A contemporary example of this process will be presented at the opening of Chapter 6.

In the 1990s, Palagio Engineering brought to market a line of back-ventilated wall tiles designed by architect Cristiano Toraldo di Francia.[16] These products grew from the company history as a high volume pro-ducer of floor tiles. Modeled on those techniques, two optimized wall tile

[16] Toraldo di Francia, p. 51.

Figure 3.19 Sample of extruded profile design for precast concrete backing. Inserted face down in the concrete panel mold after removal of the flat back plate at the defined break points. NBK North America, Salem, Massachusetts. Photo: Donald Corner.

sections were extruded back to back as a symmetrical profile, then split apart and trimmed up before shipping. Currently, this approach is used for veneer tiles that will be backed up with concrete as an integrated, precast panel (Figure 3.19). The "back" of the section stabilizes the form through production. It is cut away at predetermined weak points to produce a single finished face with a complex shape. These units are placed face down in the panel mold before the concrete is poured.

Custom milling of fired terra cotta can be done with hand tools or very precise, automated equipment. This includes boring holes for back fastening, cutting the ends for a mitered corner, or milling the ends to create lapped, horizontal joints on vertically oriented planks (see Figure 9.5.4).

Slicing across complex, extruded profiles can produce a rich, sculptural outcome, as demonstrated by the Wave/Cave designed by SHoP Architects (2017). The open cores were appropriate for a temporary exhibition at Fuori Milan, 2017, in Italy. For a long-term building application, compound angles can be milled in terra cotta components to fit plugs or caps that close the tops of clustered vertical extrusions. These parts may be glued together, with only the glazed, outer surfaces showing. The alternative is fully integrated closure pieces, custom formed by other means. The first strategy was applied at the top of the fluted panels on 111 West 57th Street, NYC, also by SHoP Architects (see Figure 3.24). The second approach was used at The Fitzroy, featured as a case study in Chapter 9 (see Figure 9.1.7).

DEVELOPING ORNAMENT

Ornament (noun): a detail used to beautify the appearance of something to which it is added or of which it is a part.[17]

17 www.dictionary.com, accessed April 6, 2021.

This definition of ornament speaks directly to the possibilities of work in terra cotta. Aesthetic content can be developed in each, individual piece, as well as in the larger wholes those pieces form in aggregate. Through all of the production processes just described, there is a means to deliver discrete sculptural forms that are explicitly ornamental. These can be elegant

Figure 3.20 Above: Rehabilitated Utah State Capitol Building, Salt Lake City, UT. Photo: Brett Drury. Below: Full scale mockup of replacement parts at Boston Valley Terra Cotta, Orchard Park, New York. Photos: Donald Corner.

sectional profiles realized by extrusion, or complex patterns rendered by pressing.

A significant part of the formal language of architectural terra cotta derives from carved stone, the material it was called upon to replace. Terra cotta shares, to a sufficient degree, the precision of stone and the ability to make complex assemblies of individually complex parts. Perhaps the most vivid example is the replication of classical capitals in terra cotta for banks, state houses and other civic structures, all across the United States. These appear in the iconic, black and white photographs taken by Mary Swisher at Gladding McBean.[18] Boston Valley Terra Cotta continues making replacement parts for these structures, as they did for the Utah state capitol (Figure 3.20).

The search for an appropriate architecture of tall office buildings weaves through the chapters of this book, as the time period of this development coincided with the golden age of American terra cotta production. It was a marriage driven by building function and construction technique, but with a large measure of aesthetic or stylistic interest. The great debates about ornament, its origins in nature or in academic traditions, and its place in a wholly new building type, were played out in clay. Terra cotta was an effective medium for each manner of expression.

Virginia Guest Ferriday traces then predominant Beaux-Arts design principles across the face of early twentieth-century buildings in Portland, Oregon. A proper building should have a base, attic, shaft and cap.[19] For some time, the repetitive floors of the shaft remained punched windows in brick cladding. The base was opened up with storefront glazing. However, the horizontal boundaries and transitions at the attic and cap were the ideal locus of classical components in terra cotta. Each of them might be several stories tall, composed with a language of supports (columns and pilasters), bands (friezes and cornices), sculptural panels, and applied, free-standing elements (Figure 3.21).[20]

Louis Sullivan, in "The Tall Office Building Artistically Reconsidered" concurred with a tripartite division of the building, but for reasons of function: special commercial uses on the first one or two stories, offices stacked layer upon layer in the middle, and a cap presumed to be unoccupied technical space. He disavowed ornamental programs that made the buildings "a field for the display of architectural knowledge in the encyclopedic sense."[21] Instead, he argued for the essence of the thing to take its own shape, as things do in nature. The essence of a tall building was to be "lofty"; "a proud and soaring thing rising in sheer exultation that from bottom to top it is a unit without a single dissenting line."[22]

[18] Kurutz, p. 34.
[19] Ferriday, p. 20.
[20] Ferriday, p. 27.
[21] Sullivan, 1896, p. 34.
[22] Sullivan, 1896, p. 33.

Figure 3.21 Wilcox Building, Portland, Oregon (1911). Whidden and Lewis, Architects. A twelve-story, steel frame, commercial building, with the upper two floors composed as a frieze below a large, overhanging cornice. Executed in white terra cotta as the metaphorical capital, above a column shaft of white brick, now soiled and resistant to cleaning. Photo: Michael Yauk.

Figure 3.22 Louis Sullivan's Guaranty Building, Buffalo, New York (1895). Right: Detail of the ornament at the Church Street entrance. Photos: Donald Corner.

The Guaranty Building in Buffalo, NY (1895) is clad entirely in red terra cotta. The internal functions are clearly delineated, and the verticality of the frame structure is powerfully expressed. The surfaces are humanized with an ornamental program sketched by Sullivan and translated into three dimensional molds by the sculptor Christian Schneider at Northwest Terra Cotta Company in Chicago (Figure 3.22).[23] Boston Valley Terra Cotta reproduced many intricate parts at the base of the building when it was recovered from unfortunate alterations.

A great deal has been published about the Guaranty Building and Sullivan's concepts of ornament, beginning with Sullivan himself (*A System of Architectural Ornament According with a Philosophy of Man's Powers*, 1924). What can be read directly in the details is a seamless fit between his intentions and the processes for achieving them in clay. Much of the building surface is elaborated with geometric details generated by subtraction. The surface of the original model has been carved away, as one

[23] Guaranty Building Interpretive Center.

Figure 3.23 Guaranty Building, Buffalo, New York. Above: Repeating ornamental motifs at the central portion of the elevation. Below: Incised and additive forms at the column capitals near the base. Photos: Donald Corner.

might gouge or chip carve for a block print. These incised portions are framed by hardline figures that speak of drafting tools: the compass, divider, triangle and straightedge. At strategic points, the organic elements literally grow out of the surface by engaging clay sculpture as an additive process. This duality allows the building surface to be read as both richly developed and tightly aligned with the structural form underneath. The Guaranty is a model for contemporary work in terra cotta; as it speaks of the repetitive, constructive aspect of cladding while embracing ornamental opportunities that derive, efficiently, from the means chosen to form the parts (Figure 3.23).

As a visitor to the University of Oregon, architect Nicholas Ault presented a pedagogy in which students of design unpack the geometric operations used to generate Sullivan's ornament. They do this to understand the generative power of drawing, and subsequently make a transfer of thought and process to the digital tools now in their hands.[24] Terra cotta

[24] Ault, pp. 316–20.

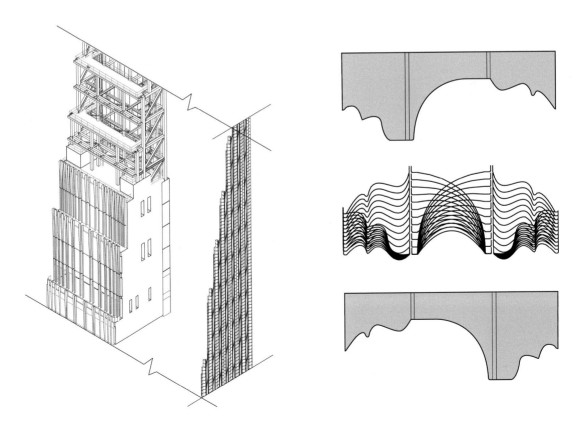

Figure 3.24 Stepped profile of the residential tower at 111 West 57th Street, New York, NY (2020). Vertical pilasters of extruded terra cotta transform continuously, with the profiles alternating back and forth through stages generated by parametric design. Images courtesy SHoP Architects and JDS Development Group.

remains a perfect medium for constructing formal transformations developed through parametric design. These can apply to the overall form of the piece as well as the surface. Shapes conceived in the digital realm can be applied directly to the automated cutting of an extrusion die, or the fabrication of a first positive around which the molds are made. Analogous to the Guaranty Building, blocks of modeling material can be carved away with routers and water jets, or built up with 3-D printing. The project designed by SHoP Architects, at 111 West 57th Street in New York, includes a parametric transformation of fluted pilasters designed to capture sun and shadow effects on the east and west elevations of the tower (Figure 3.24). The ornamental strategies developed for admittedly high-end projects will find their way into the mainstream as a marriage of digital conceptions and the versatility of clay.

CHAPTER 4

Color, Texture and Finish

For many, the first thought of terra cotta is of roof tiles, unglazed and red. Architectural theorists have, at various times, associated the soft, red surface with the true nature of the material (Figure 4.1). By contrast, advocates within the industry enthusiastically market a broad range of colors and textures impossible to match with any other single material.[1] Concepts of terra cotta use have cycled between reductive phases of aesthetic theory and expansive phases of bold experiment.

[1] White, p. 6.

Figure 4.1 Red terra cotta without glaze at Louis Sullivan's Guaranty Building, Buffalo, New York (1895). Photo: Donald Corner.

DOI: 10.4324/9780429057915-5

The creative explosions in each cycle demonstrates that easy access to extraordinary variety is a central, defining characteristic of architectural terra cotta.

COLOR OF THE CLAY BODY

Color options begin with the clay chosen for the structural body of a piece, independent of surface treatments or applied finishes. "Through body" colors have an undeniable appeal for a material taken directly out of the ground. Clays chosen for performance attributes become associated with their color. The red roof tiles covering Brunelleschi's iconic dome at Santa Maria del Fiore, in Florence, are a signature of broadly influential Tuscan architecture.

On the plains of Lombardy, southeast of Milan, there was very little building stone.[2] Burned clay was the available material for the structural mass of the building and for the surface ornament. The churches of Pavia and Crema, spanning from the late thirteenth into the fifteenth century, show a continuous transition from brick, to carved brick, to unglazed terra cotta. Santa Maria del Carmine, in Pavia, contains all of these materials in a closely integrated Lombard Gothic façade that is consistent in its color and character. It was this region that Lewis Gruner and his team documented in 1867.[3] Their careful drawings and color plates were a reference for brick and terra cotta combinations that continued into the early twentieth century (Figure 4.2).

Influence of the Italian Renaissance carried terra cotta techniques across Europe, including the polychrome glazes of the Della Robbias. That influence came again with the revival of Palladian architecture. Unglazed terra cotta appeared as ornamental elements: string courses, cornices, balustrades and decorative panels, in replacement of carved stone. In the United Kingdom, and subsequently in the United States, there was a clear choice to be made as to whether these pieces should contrast or complement the dominant brickwork. The complementary option, particularly in the United States, usually implied dark red, unglazed units.[4]

The preference of designers was color that could be produced from local clays. In New York, prior to the 1880s, red was assumed to be the only choice. Natural clays, without mineral additives, result in white, buff or red terra cotta. Gray terra cotta soon followed, and these colors dominated the decade.[5] In Chicago, much of the terra cotta was a "grayish buff" that resulted from a clay dug in Brazil, Indiana. This was favored because it resembled nearby Joliet Limestone.

In New York, an early adopter was George G. Post. The Long Island Historical Society Building (1879), had ornamental elements made of a red/

[2] Gruner, p. 3.
[3] Gruner.
[4] Prudon, Abstract.
[5] Elliot, p. 56.

Figure 4.2 Rose window of carved brick and terra cotta at Santa Maria del Carmine, Pavia, Italy (1374–1461). Photo: Donald Corner.

orange, unglazed terra cotta to complement the body of dark red brick.[6] Post went on to do a series of taller structures: Smith Building (1880), Post Building (1881) and Mills Building (1883). He developed an aesthetic system in which the ornamental elements were expressive, but subservient to the structure. On the Mills Building he chose unglazed, red terra cotta for recessed spandrels, capitals, cornices and parapets to differentiate them from the load bearing piers of brick and stone.[7]

As the full skeletal frame took over, the benchmark for integrated façade design was the Wainwright Building in St. Louis (1891) by Adler and Sullivan. The base is clad in brownstone, the verticals of the central shaft still of brick, and the spandrels and uniquely ornamented attic are molded in unglazed terra cotta. The three materials accentuate the functional differences among parts that remain compatible in color and texture.[8] The Rand McNally Building in Chicago (1891), by Burnham and Root was the first with a structural frame entirely of steel. It was also the first to eliminate brick and be clad completely with unglazed terra cotta.[9] Rolling forward nearly a full century, terra cotta

[6] Hamrick, p. 18.
[7] Weisman, pp. 183–9.
[8] Hamrick, p. 26.
[9] Tunick, p. 15.

rainscreen cladding picks up the threads of this evolution, expressing a lightweight, non-bearing skin. This functionalism is humanized by the repeating module, color and texture of the units. The "natural" choice, is unglazed clay.

Regional preferences for color developed early in the American terra cotta industry, but these did not necessarily imply local clay.[10] "The first red terra cotta in New York City was made in Chicago from Ohio clays, while the first red terra cotta building in Chicago was made in Perth Amboy from New Jersey clays."[11] Regardless of origin, clay is delivered as a finely ground powder. Contemporary facilities have automated mixing of ingredients from a varying number of hoppers or silos. Metal oxides and other chemical additives customize the body color. The degree to which the final colors are native to the clays varies from one producer to another. A facility could have as few as two base clay colors, white or red, and develop color families from there. Another facility may draw from a greater range of clay types with less dependence on additives. Either way, a great range of body colors is available, as well as subtle gradations for which the percentage of a chosen additive and the ultimate firing temperature are variables (Figure 4.3).

Clay mixes include solid aggregates, usually grog, added to control shrinkage. At 30% to 35% of the composite volume, grog can have a significant influence on both color and texture. Grog can be made by grinding up and re-using the cut offs, trimmings and rejected units of fired clay produced on the same site. The waste stream has to be sorted by color range and composition. Glazed waste may be sent to brick manufacturers where the glaze content does not interfere with color outcomes. Conversely, rejected bricks can be ground up as grog when a particular blend or accent is needed in the terra cotta.

For units with a closed, smooth finish, the color of the "fireskin" will derive from the clay itself, rather than the aggregates. Various textures expose the aggregates, and the relative sizes of the particles affects their visibility. Finally, a fired clay unit can be sand blasted or surface ground to remove the skin and make the particles in the body readily apparent (Figure 4.4).

TEXTURE OF THE SURFACE

Texture can be developed at a range of scales. Mixing contrasting units in a pattern creates texture at the scale of the building. Salt glazes and other special effects create textures at the scale of the finish. This section will focus on texture within a single unit, using techniques applied to the wet clay immediately after it has been formed.

[10] Tunick, p. 12.
[11] Stratton, p. 29.

Figure 4.3 Single fired terra cotta test samples at Palagio Engineering. Left to right shows effect of increasing the firing temperature. Top to bottom shows greater percentages of metal oxide colorant. Photo: Donald Corner.

Figure 4.4 Above: Finish sample sandblasted after firing to expose aggregate in the body, then coated with a clear glaze. Below: Salt and pepper finish created by grinding and polishing the sample. Palagio Engineering, Greve in Chianti (Florence) Italy. Photos: Donald Corner.

Hand pressed terra cotta requires a high level of manual finishing, to clean and sharpen the ornament, as well as fill and smooth voids on the surface. As an economical substitute for stone, terra cotta often simulated stone finishes, including chisel drafted margins and furrowed surfaces (Figure 4.5). While still craft work, combed or raked textures can be applied much more efficiently to wet clay.

Terra cotta adapted readily to the Art Deco style in the late 1920s. As production shifted toward ceramic veneer, the ornamental reliefs were flattened and simplified. Sculptural panels were limited to key locations, while the bulk of the façade was covered with repeating units (Figure 4.6). Contemporary terra cotta is predominantly formed by extrusion, and manual intervention is held to a minimum. The common techniques for adding texture are those that can be applied within an automated context. As the clay column emerges from the extrusion die, it can be scratched or combed lengthwise along the surface. The density of the teeth in the comb can be varied for a fine or medium effect. Grooved patterns at a larger spacing can be incorporated into the die.

These tools tear the skin of the extruded clay, leaving ragged edges at a small scale. The degree of roughness is influenced by the density and size of aggregate particles incorporated into the mix. Larger particles will be raked to the surface, exposing them as a change in color as well as grain. This is particularly important for a "wire struck" finish in which a taut wire peels off the top layer of fine clay. Aggregate particles dragged along the surface by the cutting wire produce the final texture. The die is modified to increase the initial thickness of a clay face that will be cut down. Such textures are usually applied across the full surface of flat planks that are intended to be left unglazed. However, they can be developed on a selected portion of a profile, glazed or not (Figure 4.7, top image). At a 2019 workshop sponsored by Boston Valley, a team of architects from Kohn Pederson Fox prototyped custom rollers that press textures into a bold, extruded profile. Both the extrusion and the roller were designed with digital tools for an exact match of the curvatures and to establish the pattern on the roller face that would produce the desired result on the extruded clay (Figure 4.7, bottom image).

As responsibility for texture is transferred from scratching tools to the die shape, texture grows from a micro to macro scale. Fluted surfaces can be fine grained, bolder, or the two combined. The profiles developed for Rogers Stirk Harbour at the Scandicci town center are deeply undercut to produce texture through the contrast between color and shadow (see Figure 6.34).

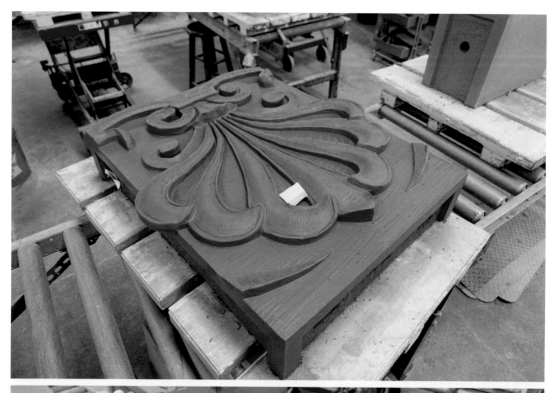

Figure 4.5 Above: Hand pressed terra cotta block with fine grained surface textures applied in the finishing area. Below: Terra cotta block simulating tradition stone margins and textures. Boston Valley Terra Cotta, Orchard Park, New York. Photos: Donald Corner.

Figure 4.6 Textured blocks combined with sculptural relief at the Wiltern Theatre Building, Los Angeles, California (1931). Designed by G. Albert Lansburgh. Photo: Donald Corner.

APPLIED FINISHES

The application of glaze to architectural terra cotta opens a final means of expression with virtually unlimited range.

> Glaze may be defined as a continuous adherent layer of glass, or glass crystals, on the ceramic body. The glaze is usually applied as a suspension of the glaze-forming ingredients in water, which dries on the surface of the piece in a layer. On firing, the ingredients react and melt to form a thin layer of glass. The glaze may be fired at the same time as the body or in a second firing.
>
> The main purpose of the glaze is to provide a surface that is hard, nonabsorbent, and easily cleaned. At the same time the glaze permits the attainment of a greater variety of surface colors and textures than would be possible with the body alone.[12]

[12] Norton, p. 171.

Figure 4.7 Above: Extruded profile that includes fine surface texture. NBK Keramik. Below: Prototype for texture applied to a curving surface with a roller. Developed for Architectural Ceramic Assemblies Workshop (2019) by a team from Kohn Pederson Fox Associates, with technical support from Boston Valley Terra Cotta. Photos: Donald Corner.

Glaze is an ancient technology found on bricks, tiles and tableware. Precedents extend from the Middle East, across to China and back into Europe through the Moorish culture in Spain. Babylonians and Assyrians applied colored glazes to bricks and wall tiles from the ninth to the sixth centuries BCE.[13] Persian craftsmen maintained the techniques of the Babylonians and the Egyptians before them. They also traded with the Chinese, providing a link from the east to west in ceramic products. The Chinese developed glazed porcelaneous stoneware from the fourth to third centuries BCE. They used kaolin, a pure white material subsequently referred to as "China clay."

Western architectural traditions come through the Greeks and Romans who made extensive use of architectural terra cotta, but without glaze.[14] The technology re-entered Europe with the invasion of Spain by the Moors (710 CE) and the trading of Islamic pottery and tiles into central Europe. Italian craftsmen developed tin-enameled majolica ware, which in name derives from Majorca and the Hispano-Moorish thread. Centers of production developed in Florence, Rome, Venice and Faenza.[15] It is from the latter that the French term "faience" derived, still used in the United Kingdom to describe glazed terra cotta.

The Italian masters were the Della Robbia family of Florence: Lucca (1400–82), his nephew Andrea (1435–1526) and Andrea's sons Girolamo, Lucca and Giovanni. Lucca della Robbia's Madonna and Child (Figure 4.8) is thought to be one of the pieces to adorn Brunelleschi's landmark Foundling Hospital when it opened to orphans in 1444.[16] Andrea's famous tondos of babes in swaddling clothes were placed on the façade in 1487. Work by the Della Robbias is a significant benchmark in the history of glazed architectural terra cotta. Their glaze formula was, by legend, a closely guarded family secret. The reputed secrecy foreshadows a long tradition of proprietary control over glaze technologies that have been developed through a mixture of science and art.

High gloss ornament declined in the Italian Baroque period in favor of the many things that could be done with stucco. The Florentine sculptor Giovanni da Maiano was commissioned to prepare decorative terra cotta rondels for Hampton Court Palace, Surrey, dated to the 1520s.[17] In a letter, Da Maiano describes them as painted and gilded, rather than glazed. Nearly three centuries later, Coade Stone took over, formulated to present a granular surface that resembled stone. With glass as an internal aggregate, Coade Stone did not need to be glazed.[18]

Before 1840, the few terra cotta elements used in the United States were imported from England, although by 1849 there was domestic production

[13] Davey, p. 69.
[14] Davey, p. 69.
[15] Davey, p. 89.
[16] Museo degli Innocenti.
[17] Henry et al., p. 787.
[18] Henry et al., p. 669.

Figure 4.8 Madonna and Child, Lucca della Robbia (1445–50). Glazed terra cotta façade element displayed at the Ospedale degli Innocenti, Florence, Italy. Photo: Donald Corner.

in Massachusetts. The architecture of James Renwick and Richard Upjohn introduced terra cotta elements to New York City in the 1850s. Ornamental cornices and animal heads on the window surrounds were also painted to more nearly resemble stone.[19]

The potential for terra cotta to be an independent, directly expressed, polychromatic material was recovered step by step in the latter half of the nineteenth century. Theorists argued that the softer character of natural clay should be featured, and sculptural modeling should not try to compete with the crispness of cut stone.[20] With the promise of lower cost and better resistance to the harsh, soot filled environment of the Victorian city, the material won acceptance on major commissions. Seminal buildings included New Alleyn's College, Dulwich (1866–7) by Charles Barry, Jr., the largest terra cotta application of its time.[21]

Prince Albert's cultural center in South Kensington featured more landmark buildings. The core of what is now the Victoria and Albert Museum (1867) by Sykes, Fowke and Scott used pale cream terra cotta for sculpture and applied ornament, contrasting with smooth red brick.[22] The Natural History Museum (1873–81), by Alfred Waterhouse, has extensive façades entirely of terra cotta covering the brick cores of the walls and the embedded iron structure.[23] Waterhouse was fully convinced of the pragmatic advantages of terra cotta in the urban environment and he shifted the project to a Medieval language to embrace the subtle color variations that were inevitable from one block to the next.[24] The building is also significant for the rich ornamental program of plants and animals, naturalistic forms popularized by Owen Jones and John Ruskin.[25] The writings of Ruskin and others are credited for giving terra cotta an appropriate place in the theoretical debates of the age.[26]

The path to polychromatic terra cotta in the United States tracked developments from England in the early years, before branching out in new directions. From the 1850s, there were adoptions in New York, Chicago and Boston.[27] In 1871, winners of the Museum of Fine Arts competition in Boston were announced. The submission by John H. Sturges and Charles Brigham featured terra cotta from the start, and the building would go on to become a significant waypoint in the growth of terra cotta in North America. Buff colored terra cotta details, contrasting with broad expanses of red brick, closely followed the English precedents, as did the use of natural forms in the ornament, advocated by Ruskin. Also attributed to Ruskin, the scale of the ornament became progressively larger and more abstract, moving up the building, farther away from the viewer.[28]

When the first phase of the museum was completed in 1876, American architecture was shaped by a wildly eclectic array of stylistic influences.[29]

[19] Prudon, pp. 16–17.
[20] Elliot, p. 56.
[21] Hamrick, p. 8.
[22] Hamrick, p. 9.
[23] Henry et al., p. 676.
[24] Elliot, p. 58.
[25] Elliot, p. 58.
[26] Hamrick, p. 8.
[27] Hamrick, p.11.
[28] Hamrick, p. 12.
[29] Hamrick, p. 13.

Despite the opportunity to break new ground, a more secure route was to use terra cotta as a replacement for stone, working within established traditions.

When the explosion of color and finish finally came, it was attributed to the work of architect Stanford White and the Columbian Exposition of 1893. Stanford White's appreciation of ornament lead him to expand on the glazes specified for both terra cotta and brick.

> White worked in close collaboration with the Perth Amboy Brick Company to develop a wider range of colors – first gray, then white and buff, as well as various mottled and speckled glazes for both brick and matching terra cotta pieces.[30]

When the Columbian Exposition opened in 1893, it became known as "The White City of Chicago." The architects of the fair, and those inspired by it, opened a new era of urban architecture, with classical detailing in bright, white glazed terra cotta reflecting light into the city streets. The Reliance Building (1894) in Chicago, by D.H. Burnham, was the first skyscraper to be clad in clay units with a cream-white glaze. Literally hundreds of tall white buildings followed.[31] San Francisco by the late 1890s gave up on the conservative earth tones, switching to light values with a variety of colors and glazed surfaces as the chemists at the Gladding McBean Company advanced their craft.[32]

In the early years of the twentieth century, the streets in the commercial core of Portland, Oregon were formed by nine-story buildings, with over-hanging cornices to complete the sense of enclosure. Terra cotta façades were light in color and highly reflective. Integrated light bulbs outlined entrances and building tops so that the district sparkled by day and by night. The atmosphere is described as having resembled the 1893 Chicago Fair and Portland's own 1905 Lewis and Clark Exposition (Figure 4.9).[33]

When English potters, T.C. Booth and his father, came to the United States, they intended to make majolica art objects. Booth convinced William Hall, of the Perth Amboy Company, in New Jersey, that he could implement a process applying glaze directly to green clay and firing it in a single kiln cycle.[34] This method is dated from 1888 in British histories.[35] Previously it had been thought that glaze would only succeed when applied to a biscuit fired body, with the whole fired again. Single firing was a significant savings in cost and delivery time.

Booth developed a full glaze in a creamy white that was of great interest to architects Stanford White and Bruce Price. They wanted a dull white finish that would resemble the texture of marble. Booth achieved this by applying heavy coats of full glaze and then sandblasting the surface to

[30] Roth, p. 200, note 24.
[31] Tunick, p. 15.
[32] Kurutz, p. 94.
[33] Ferriday, pp. 15–16.
[34] Geer, p. 208.
[35] Henry et al., p. 649.

Figure 4.9 Window detail, Meier and Frank Building, Portland Oregon. Designed by A.E. Doyle in 1910 and executed in phases. Photo: Donald Corner.

produce a matt finish.[36] At the close of the nineteenth century, Beaux-Arts classicism was the fashion for larger, public buildings. Terra cotta of light color was favored in combination with light colored stone. Because the terra cotta was applied to the middle and top of taller structures, it was often difficult to distinguish between the two materials.[37]

McKim, Mead and White were credited with the introduction of polychrome, glazed terra cotta to American architecture, with the design of the Madison Square Presbyterian Church.[38]

> The front was a triumph of restrained color. The blue in the pediment, the white of the angels, the dull gold tops of the pillars, with blue behind, the splendid granite shafts of the columns, of a gray that was almost green, the apple-green overlaid with gold in the line of the eaves, the yellow and cream – yet all so quiet, so harmonious, so unobtrusive.[39]

Over the same time period, glazes with strong color and texture had emerged in England. Doulton and Company supplied material for the

[36] Geer, p. 209.
[37] Prudon, p. 44.
[38] Tunick, p. 17.
[39] Geer, p. 212.

Royal Arcade, Norwich, in 1899, and the Anchor Pub, London in 1898. In 1907, Leeds Fireclay Company provided a saturated ox-blood red cladding for the Tufnell Park underground station, and strong colors were used to identify stations on the various city lines.[40]

In California, the catalyst for a new era in finishes was the 1906 San Francisco earthquake. The resulting fire destroyed many buildings that had terra cotta over iron and steel structure. Among those lost was Gladding McBean's own headquarters on Market Street. Post-disaster code revisions increased the thickness of fire covering, and rapid reconstruction of the city was an overall boost to the industry. A landmark was the Hearst Building (1909), with pink Tennessee marble on the base, white glazed terra cotta in the middle, and a polychrome glazed top. The structure and interior partitions dropped the use of terra cotta in favor of reinforced concrete and plaster. As concrete construction expanded, architects turned to a much more expressive use of terra cotta as the exterior cladding to relieve what they perceived as potentially dull façades.[41]

Joseph Baldwin DeGolyer came to Gladding McBean in 1888 as a civil engineer with specialty training in chemistry. He expanded the "architectural department" from one kiln to eighteen kilns and 300 men by the 1920s. His chemistry background allowed him to keep pace with new developments in polychrome finishes. In the post-earthquake years, he added dozens of new colors to the Gladding McBean catalogue.[42] Many architects, with aesthetic systems based on stone, preferred the matte finishes, similar in texture to unpolished marble. The full glazes were considered by some to be too bright for the larger new buildings. Gladding McBean, like the rest of the industry, offered a range of earth tones, with the ornamental elements developed in greens, reds, blues, whites, yellows and golds.[43]

Stimulated by the earthquake recovery, a golden age for terra cotta opened on the West Coast of the United States. Buildings influenced by the Beaux Arts Schools featured ornamental motifs that exploited the plasticity and color palette that terra cotta could bring to the work (Figure 4.10).[44] Terra cotta was a favorite for theaters and movie houses on both coasts. Rich color and imagery suited the glamor and fantasy of these venues.[45] San Francisco architect, G. Albert Lansburg, designed the New Orpheum Theater (Palace Theater) in Los Angeles in 1911, using strong color inside and out. There were allegorical figures supported by polychrome cornices, friezes, mascarons and other three-dimensional ornament. The Orpheum was described as the first colored façade to be built in Los Angeles, and one of the first in the West. Along with other commissions of the period, it foreshadowed the explosion of polychrome work in the 1920s.[46]

[40] Henry et al., p. 679.
[41] Kurutz, p. 95.
[42] Kurutz, p. 95.
[43] Kurutz, p. 96.
[44] Kurutz, p. 96.
[45] Tunick, p. 24.
[46] Kurutz, p. 120.

Figure 4.10 Commercial structures on Broadway in Oakland, California. Above: The Cathedral Building, by Benjamin Geer McDougall, Architect (1914). Below: Roos Brothers Building by William Knowles, Architect (1922). Photos: Donald Corner.

The 1930s brought dramatic changes due to a confluence of factors. Large extruded blocks of terra cotta were cut to form ceramic veneer panels of much broader dimension than their hand packed predecessors. The discipline of the extrusion process produced more subtle surface articulation. At this same time, the science of glazes reached a level of maturity that provided a bold range of colors and textures to enliven the generally flatter façades. Third was the coincident shift in architectural style, favoring the streamlined character of the Art Deco or the Art Moderne. Henry Russell Hitchcock and Phillip Johnson, while defining the International Style, decried the past history of terra cotta blocks as unsatisfactory because they were too suggestive of traditional masonry, and presumably out of step with the new age (Figure 4.11).[47]

The landmark of this era was Raymond Hood's McGraw Hill Building of 1931. It was the last of a series of towers he designed in New York, articulated at the top with step backs and terra cotta finials intended to provide a strong silhouette against the sky, day or night.[48] It was also the first major project to be clad in machine made terra cotta.[49] The chosen color was difficult to manage at the time, emerging from the kilns with a high degree of variation. By account, the blocks were blended to mitigate the contrasts, their placement having been directed by staff from Hood's office, perched in an office space across from the construction site.[50]

Looking back now on the long struggle to control the colors of terra cotta, we might caution against too much success. Subtle variations in color are fundamental to the richness of the material. If complete uniformity is a primary objective, it opens the door to alternative materials that are far less expensive than terra cotta.

A full account of glaze technology is beyond the scope of this book. On factory visits to Gladding McBean, the recommended reference was Tony Hansen's classic, *The Magic of Fire: Fighting the Dragon with Insight and Foresight: Understanding the Chemistry and Physics of Glazes, Clays and Materials*. It has been superseded by extensive, interlinked websites by Digital Fire Corporation at digitalfire.com. These references explore the materials used in clay bodies, and in glazes, describing their physical, chemical and mineralogical properties and the means to control them. Other outstanding sources are online courses available through the New York State College of Ceramics at Alfred University. That faculty collaborates with Boston Valley Terra Cotta, also in western New York.

With glazes, the ingredients, application techniques and firing process all influence the outcomes. The act of glazing is introduced as making

[47] Prudon, p. 46.
[48] Tunick, p. 22.
[49] Elliot, p. 58.
[50] Tunick, pp. 23–4.

Figure 4.11 Los Angeles Jewelry Center (1930–1). Art Deco landmark designed by Claud Beelman as the Sun Realty Building, 629 S. Hill Street, Los Angeles, California. Above: Upper stories. Below: Detail at the mid-level. Photos: Donald Corner.

glass and making it stick to the surface of a clay body.[51] The many materials that can be used to make a glaze fall into three major groups, with one additional component.

Glass Formers

Most often this is silica (silicon dioxide, SiO_2), best found in quartz or flint. It requires very high melt temperatures, over 1300°C, were it not for other ingredients.

Stabilizers

Molten glass would run off the work piece if it were not stabilized to control the viscosity. Alumina (aluminum oxide, Al_2O_3) is used to do this. It stiffens the melted glass to control the flow. Aluminum oxide also improves the resistance of the glaze to cuts and scratches.

Melters

Fluxes are used to lower the melt temperature at which the glass is formed. Different fluxes operate effectively in different temperature ranges. They include oxides of the alkali metals (Lithium, Sodium, Potassium) and of the alkaline earths (Magnesium, Strontium, Barium). Zinc oxide and lead oxide are also effective melters, although lead has lost its historic significance due to the toxicity. Zinc oxide is used in what are known as the Bristol glazes.[52]

Clay

The final ingredient is clay itself: kaolin, ball clay or bentonite. Glaze mixtures are not solutions, but aqueous suspensions. Clay slows the tendency of the other ingredients to settle toward the bottom of the mix bucket. It also helps bond the mix to the clay body so that it does not flake off before firing in the kiln.

Beyond the base formula, there are additives used in small amounts to adjust the properties of a glaze. The objective is to control the glass, in both its appearance and performance. The designer is likely to focus on the first, relying on the glaze master for the second. Foremost among the additives are the colorants; metal oxides or metal carbonates, used in combinations to develop specific hues. The development of colors is a patient art because of the many variables involved. Unlike paint, the appearance of the mix gives no direct hint as to the final outcome. Clay samples must be prepared, glazed, fired, evaluated and systematically recorded before

[51] Blue Mountains Creative Arts Center.
[52] Norton, p. 180.

anything is known. Colors can be subtly influenced by the percentage of oxides used and the selected firing temperature (see Figure 4.3).

Glass wants to be clear, and that is desirable if the underlying clay body has a special color or texture intended to be seen in the final result (see Figure 4.4). The clear glaze closes the surface, making it easier to clean and maintain. When the clay body is not to be seen, opacifiers are added to the glaze chemistry. By long tradition, tin is used to introduce un-melted particles that are suspended in the glass, reflecting back the penetrating light. Certain ingredients release gaseous bubbles as the glass forms, with a similar blocking effect. A deep and potentially interesting form of opacity may result from phase separation, through which the glass melt divides into two or more liquid layers that have a different chemistry and a different appearance.[53]

Glaze formulas are most likely to produce a glossy appearance. It takes extra effort to produce semi-gloss or matte finishes. A matte finish results from micro-crystals forming on the surface of the glass during cooling. Their formation can be predicted on a Stull Chart, plotting the relative content of silica and alumina on the axes of a graph. Most matt glazes result from a high alumina content.[54]

Perhaps the most significant requirement is control of the coefficient of expansion. If the bond between a glaze and clay body is to remain intact, they must have good "fit." They must be sufficiently close in their thermal expansion and contraction. Glass and glazes are weak in tension and strong in compression. The objectives for glaze are analogous to those for tempered or heat strengthened glass. When the inner core of a glass sheet cools, it pulls the previously solidified surfaces into compression, increasing the bending strength. A surface glaze is also designed for built in compression. This guards against the chance that future differential movement relative to the clay body will put the glaze into tension. Too little compression and the glaze might crack or craze. With too much compression, the glaze may separate, often at the edges or corners of a shape. This is called peeling or shivering. Under certain design conditions, some amount of crazing might be inevitable. In other cases, in may be the designer's intention, as a special effect.

Terra cotta finishes can be roughly ordered in terms of the degree of intervention, or departure from the character of the untreated clay body.

The first increment of change is to finish the clay body with a clay slip. This is a slurry or water borne suspension of quality clay particles that is applied to the body before firing. The result after firing is a fine surface texture with a more even composition. It might be considered a "natural" outcome, improving the product without much change in its character.

[53] Digital Fire Corporation, "Glaze Chemistry."
[54] Digital Fire Corporation, "Matte Glaze."

A slip is not a glaze because it does not reach the melting temperatures required to form glass.

Engobes are more common than slips in contemporary work, although drawing a precise difference between them is not of primary importance to the building designer. Neither one is a glaze. Engobes are usually defined as containing less clay, with more fluxes and other "glaze-like" materials. They are also distinguished by a thicker application.[55]

Whether slip or engobe, non-glaze finishes are used to alter the hue of the clay body. Because these finishes do not form into glass, they remain opaque, with good hiding power, but they do not develop color as intensely as a glaze. Glazes flow to smooth when they melt, but engobes do not. Variations in thickness left during application will show in the result. This evidence of process may be appealing in some cases.[56]

Engobes provide façades with a lively range of gentle colors (Figure 4.12), while retaining the open surface that speaks to the history of unglazed terra cotta. Architects have requested finishes that resemble the color and granular quality of weathering steel. When combined with oxygen reduction in the kiln, engobes can be used to produce dynamic color variations (Figure 4.13). Engobes can also be used as the opaque under layer for glazes to follow. The terra cotta body might be made of a red burning clay, for structural reasons. A fine grained white engobe will cover the body, masking its color and imperfections. The sealed white surface will brighten the colors of glazes applied over it.

The next plateau in terra cotta finish is to achieve a fully melted, glass coating. As previously described, glaze can be applied to clay that has already been fired, with or without a coating of clay slip or engobe. The breakthrough at the end of the nineteenth century came when full glazes could be developed during a single firing of the clay body. Once perfected, the advantages of single firing were compelling. The loading and unloading of intermittent kilns require significant time and effort. The heating and cooling of the clay products require both time and energy. High volume production facilities now have automated handling of the pieces and continuous tunnel kilns, or roller kilns that decrease these costs. Nevertheless, a single firing in any facility offers the opportunity to save energy and reduce the carbon footprint.

It is incumbent upon the designer to recognize constraints that accompany the choice to save energy. There must be a willingness to forego those colors and surface effects that can only be achieved with multiple firings. Custom formed, three-dimensional pieces are more likely to be fired in an

[55] NBK Keramik, 2019.
[56] Digital Fire Corporation, "Creating a Non-glaze Ceramic Slip or Engobe."

Figure 4.12 Central Library and Municipal Archives, Leganes, Spain (2019). BNAssociados, Architects. Half cylinders of terra cotta represent book spines. Pigmented engobe finish by NBK Keramik. Photo: Roland Halbe.

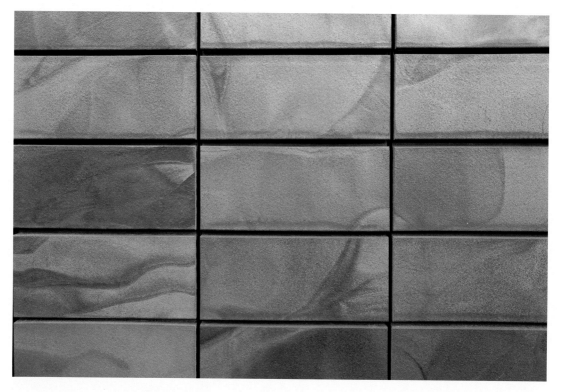

Figure 4.13 Experimental engobe finish fired in a reduced oxygen environment at NBK Keramik, Emmerich, Germany. Photo: Donald Corner.

intermittent gallery or box kiln, thus really favoring a single cycle. Relatively flat, extruded sections, such as rainscreen planks, lend themselves to automated handling and continuous kilns. However, the products are trimmed to their final length after firing, and thus the body color will be exposed at the end cuts. Depending on the level of contrast and the mounting technique, this may not be an acceptable outcome. The low-cost solution is to paint the ends. The higher quality option is a second cycle of glaze and firing.

The jump up to a double fired finish brings with it access to certain intense, fully saturated colors that may not otherwise be possible. There are special effects that can be achieved through the interaction of a second glaze layer with the first. Some of these can be approached with a single firing and others cannot. Special effects are in high demand. Fortunately, the second firing can and often must be at a lower temperature, reducing the energy penalty. Automated production lines reduce the labor penalty. The staff at NBK in Germany estimates that their made to order work is 30% single fired and 70% double.[57]

The techniques used to apply glaze produce another set of final effects. Glazes can be brushed, poured, dipped or sprayed, with spraying

[57] NBK Keramik, 2019.

the common, commercial approach since the beginning of the twentieth century. The aqueous suspension of glaze is atomized in a spray head using compressed air, as with paint. Tiny droplets strike the surface of the piece and spread out. The rate of build-up must be slow enough to allow absorption or evaporation to limit the accumulation of water. Too much water, and the whole will slough off. On the other hand, if a smooth finish is desired, the application must be fast enough that subsequent droplets merge with the previous ones into a single, thin layer.[58]

During the boom years of the 1920s, Gladding McBean developed a signature finish named "Granitex" which allowed the terra cotta cladding of tall buildings to match the color and texture of true stone used at the base. It was achieved with a specialized sprayer, called a "sputterer," creating the necessary speckles of black over gray and white.[59] They also marketed "Pulsichrome," with several colors mixed in a mottled finish. The spray head had multiple nozzles, each fed with a different glaze mix. Colors were applied with independent pulsations, or squirts (Figure 4.14).[60]

[58] Ferriday, p. 37.
[59] Kurutz, p. 124.
[60] Kurutz, p. 114.

Figure 4.14 Glaze processes at Gladding McBean Company. Left: Triple spray head for mottled application of multiple colors. Right: Three dimensional build-up of glaze on the surface. Photos: Donald Corner.

The common objective with a glaze application is to develop an economical thickness that will melt and flow together, filling any bubbles or "pin-holes" to make a uniform surface. If the clay slurry is drier and thicker, it is possible to build up differential thicknesses that will remain as a finished texture, even after the melt. A relatively subtle version of this effect is known as an "orange peel surface." It can be produced by accident, by spraying from too great a distance, or deliberately to conceal deficiencies in the accuracy of the clay body underneath.[61]

At Palagio Engineering, properties of the classic red burning clay of the Impruneta region produce an exceptionally smooth surface directly out of the extruder. Glazes are then applied with a robot in an automated spray booth. The results can be so nearly perfect that it risks defying the nature of terra cotta. The architects at Herzog DeMeuron considered it too perfect for their "Museum +" project in Kowloon. They wanted a more variegated application of green glaze that would change with the sun angles and recall the ancient clay roof tiles of China. The solution was to re-program the spray robot for less perfection so that one can see through the glaze to get hints of the body color underneath (Figure 4.15, top image).[62]

In the creative arts, there are no failed outcomes, just new process opportunities. Results that were long considered unacceptable may become an aesthetic interest. So it is now with the viscosity, or stickiness, of glazes. Chemists adjust the fluxes and stabilizers in a glaze recipe so that the molten glass layer will not run off the ware. This is a particular concern for large pieces standing upright in the kiln. Now, extruded profiles pass through the kiln face up, under conditions that are easier to control. During firing, the glaze may be allowed to flow off the ridges and puddle in the valleys, thinning the applied color at the high points and intensifying it down below. The variation in color amplifies the changes in form (Figure 4.15, bottom image).

With more stiffness, and less flow, the effects of gravity can be more subtle; gently changing the blend of colors on a convex shape (Figure 4.16, top image). With double firing, the entire surface can still retain the protective benefits of some glaze. This is not to be confused with classic "crawling" in a single fired application, where excessive shrinkage or poor adhesion cause the glaze to pull back from a crack and leave the clay body completely exposed.

Glaze recipes are generally designed for "fit" to prevent or reduce crazing. However, contemporary projects have been designed for intentional crazing, using a pattern of fine cracks as a texture. Pin-holes are another problem that can be used creatively. Unresolved specs of color, or uneven color distributions in the glaze may be failures in one context and a source of richness in another (Figure 4.16, bottom image).

[61] Norton, p. 182.
[62] Palagio Engineering, 2019.

Figure 4.15 Above: Deliberately uneven glaze application allows the terra cotta body color to show through. Palagio Engineering. Below: Two color glaze pooling. NBK Keramik. Photos: Donald Corner

Figure 4.16 Glaze effects at Palagio Engineering. Above: Gravity separated color layers on a convex surface. Below: Crackled versus smooth surfaces in a red glaze. Photos: Donald Corner.

Figure 4.17 Pilaster and parapet detail with metallic gold glaze. Selig Commercial Building, Los Angeles, California (1931). Arthur E. Harvey, architect. Photo: Donald Corner.

Figure 4.18 Above: Floral Depot Building, Oakland, California (1931). Albert J. Evers, architect. Below: Glaze detail at entrance lobby, 450 Sutter Street, San Francisco, California (1929). Timothy Pfleuger, architect. Photos: Donald Corner.

Figure 4.19 Contemporary metallic glaze effects by Boston Valley Terra Cotta. Above: Detail at 200 Eleventh Avenue, New York, NY. Selldorf Architects (2010). Photo: Boston Valley. Below: Pacific Gate Condominiums, San Diego, California. Kohn Pedersen Fox, Architects (2018). Photos: Vittoria Zupicich.

In a production setting, glazing is the purposeful, mechanized application of a material to a surface. It is, at its root, a universal process that invites adoption of techniques from other media: painting masks, decals, transfers, screen printing and inkjet printing. These approaches are more readily adapted to glazed tiles because of their limited size. With digital controls, broad graphic patterns can be divided across multiple units by "tiling." For larger units, the reach of the mechanical applicators must be greater, perhaps requiring the addition of a robot arm. Applying a mask or a masking agent to shape the spray deposition of glaze is an elementary technique that can be used at any scale.

Metallic glazes are a particularly distinguished tradition in architectural terra cotta. In contemporary work, glazes can be saturated with metal oxides, allowed to crystalize and subsequently polished.[63] Thin metallic coatings can be applied in a third firing.[64] The results can be iridescent or pearlescent. Experience with these techniques is important because not all produce stable or durable effects.[65]

During the Art Deco era of the 1920s, Gladding McBean produced numerous examples of tarnish resistant, metallic glazes in gold and silver. The Richfield Oil Building (1929) in Central Los Angeles was an acknowledged masterpiece, unfortunately demolished in 1968. The glaze colors were black, blue and gold. "For the gold coat, the terra cotta received a layer of finely pulverized gold particles held in suspension in a transparent glazing solution. This compound is fused on the surface of the individual blocks and forms a permanent coating of great brilliance."[66]

Spectacular examples of silver and gold remain in both Southern and Northern California (Figures 4.17 and 4.18). Glazes with true precious metals were a risky proposition. Loss of temperature control in firing larger pieces could cause the expensive ingredients to melt and run away.[67] Present-day glaze chemists can approach these dramatic effects without the direct use of precious metals (Figure 4.19).

[63] Bechthold et al., p. 84.
[64] Bechthold et al., p. 74.
[65] Lehmann, 2019.
[66] Allen.
[67] Gladding McBean.

CHAPTER 5

High Performance

Since 2000 BCE in Babylon, terra cotta has been used to raise the long-term performance of buildings.[1] The Etruscans, Greeks and Romans applied renewable weather skins in more explicit forms as they nailed slabs of terra cotta over wood and stone construction.[2] Overlapping tiles and slates have long been used to protect broad areas of the sidewall. The "mathematical tile" was an interesting variant, predictive of things to come (Figure 5.1). Also known as "geometrical" tiles, or "rebate" tiles, these terra cotta units were tapered, like shingles, so that they could overlap from top to bottom. The thickened weather face projected outward to make a plumb, rectangular surface that imitated a brick.[3] Overlapping side to side, they produced effective weather shells that could make a timber frame building look like brick construction with a running bond. At the height of their popularity, the tiles escaped a brick tax that was imposed on England by George III in 1784. Lighter, cheaper, easier to install, they provided excellent resistance to wind driven rain in coastal areas.[4]

Writing *The Story of Terra Cotta*, Walter Geer listed the advantages of the material as plasticity, durability (weathering), indestructibility (fire), lightness (dead loads), economy (replication) and coloration. These attributes were largely measured in contrast to stone.[5] Terra cotta fulfilled these standards when well made, but quality control depended on the producers. During the early years of American terra cotta, companies moved in and out of production frequently. Their manufacturing methods, design standards and detailing were the result of highly individualized experimentation.[6] In 1911, the National Terra Cotta Society was created, to promote uniform quality control standards. The Society published two volumes of exemplary details, one in 1914, and a second in 1927. Of particular interest are the changes that were made to recommended practices from the first book to the second. The revisions were based on "careful study of the

[1] Davey, p. 23.
[2] Geer, p. 17.
[3] Davey, p. 83.
[4] Henry et al., p. 839.
[5] Geer, 1920.
[6] Tunick, p. 7.

DOI: 10.4324/9780429057915-6

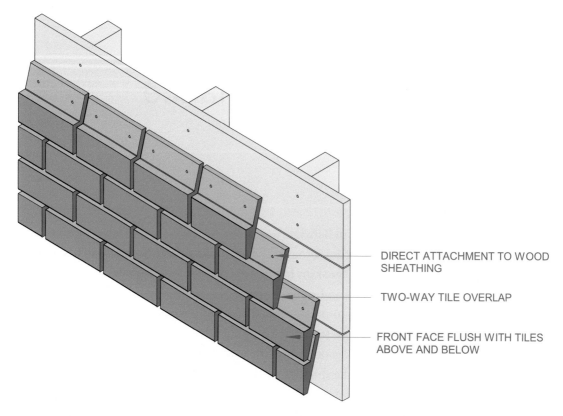

Figure 5.1 Mathematical tiles from the late eighteenth century in England. Overlapping from side to side, and top to bottom, they provide durable rainscreen cladding over wood construction. Sources: Davey and Henry.

behavior and weathering of exterior building materials."[7] They reflect fundamental principles of building science that remain in play.

First, they recommended that terra cotta facing on steel or concrete frame buildings should be continuously supported by shelf angles at each floor level. These supports were to be located in the mortar joints between courses of units. This corresponded to a major change in the building culture, from bearing walls to frame and cladding. In the United States, from the 1870s, terra cotta had been an economical means to add ornament to load bearing, masonry façades. The decorative elements were bonded into wythes of brick with little concern for support or anchoring devices. In the new building culture, terra cotta was used to cover the entire exterior surface with major roles assigned to the angles, bolts, straps and ties that held it to the frame. These ancillary parts had proven to be vulnerable. Associated with the shelf angles was the recommendation of expansion joints below each one to divide the cladding into smaller sections and isolate it from anticipated movement caused by deflection, wind load, thermal

[7] National Terra Cotta Society, Introduction.

expansion or foundation settlement. Particular note was made of the problems associated with movement in concrete frames, although frame shortening due to concrete creep was not explicitly identified. Freedom of movement between the frame and the skin was specified, along with a suggestion that it was undesirable to fill the backs of terra cotta units applied to a concrete building. By inference, this sought to avoid locking the terra cotta facing directly to the primary structure which was subject to differential movements.

The next point of emphasis was the need to protect ancillary steel components from corrosion. The prevailing practice of embedding these elements in mortar or concrete was not ruled out, but it was suggested that if the permanence of this protection could not be guaranteed, alternatives should be sought. Recommendations included corrosion resistant metallic coatings, or non-corrosive metal parts entirely.

There was a dramatic change proposed for "exposed free-standing construction." Presumably this included volumetric terra cotta assemblies not immediately backed up by a solid wall: statuary, balustrades, finials, cornices, etc. In these cases, water absorption could be expected through the mortar joints with potential damage due to freezing or the expansion of filling material inside the terra cotta. It was recommended that the internal chambers be left unfilled and ventilated with "small, inconspicuously placed weep holes."[8]

Flashings were recommended to cover large horizontal projections, protect the tops and backs of parapets, and pass underneath sill courses. Finally, all projecting elements should have laps and drips to continuously shed water.

These changes represent a fundamental understanding that despite the durability of individual terra cotta units, especially their outer faces, there is always a potential for water entry through the joints between the units or through cracks induced by structural deformation. Terra cotta assemblies should benefit from the same protections offered other forms of construction, regardless of the confidence that was held in the material itself. The mention of ventilation, weep holes, flashings, laps and drips recognized the established principles of masonry cavity wall construction and pointed toward rainscreens yet to come (Figure 5.2).

Terra cotta was (and is) integrated into many other forms of masonry construction, so it has been logical to adopt the universal mortar joint as the interface between units. Lime based mortars offered resilience that allowed the joint material to act as a cushion between the parts, while

[8] National Terra Cotta Society, Introduction.

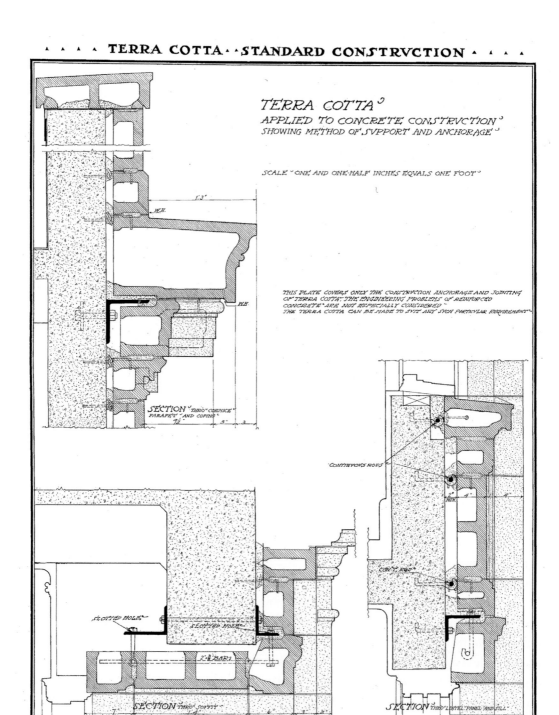

Figure 5.2 Installation detail from *Terra Cotta Standard Construction*, the revised edition, published by the National Terra Cotta Society in 1927. Indicated are laps, drips, weep holes, unfilled cores and other technical recommendations that raised the performance expectation after the 1914 edition.

maintaining a seal to keep out most of the water. The emphasis on cushioning rather than compressive strength should have become clearer as terra cotta migrated from bearing wall to cladding applications. However, the drive to improve water resistance encouraged the adoption of Portland Cement mortars that were stronger but less resilient. The literature of preservation is replete with evidence of damage to masonry units caused by mortar that is too stiff and loses its ability to cushion. These problems would have been eliminated if the joints were designed without mortar fill.

In the 1970s, the aluminum curtain wall industry propagated one of the most influential advances in building science to arise in the twentieth century, the "rainscreen principle for pressure equalized wall design."[9] During the 1980s and 1990s this concept spread across many domains of façade construction, and now it is core content in all recognized texts on building envelope performance. To summarize the principle, there should be two distinct layers to an exterior wall assembly. The inner plane should offer continuous control over airflow, because air transports most of the moisture moving through a wall. The outer plane should use drips, laps, baffles and capillary breaks to block the passage of bulk water droplets through the joints in the façade material. The magic in the assembly is the "pressure equalized" cavity between the layers, openly ventilated to match the wind induced air pressure on the outside face of the building. At the outside layer, where water is present, there is no pressure difference available to carry moisture through the joints. The inner, air barrier layer maintains the difference in barometric pressure between occupied space and the outside world. There will inevitably be pin-hole leaks in this air barrier, but they are not a threat in terms of moisture transfer if the water has already been controlled at the outer layer. In the ideal diagram of a rainscreen wall, appropriately elaborated joints in the cladding material should be left open to provide ventilation and therefore equal pressure inside the cavity (Figure 5.3).

To exclude wind driven water, the true rainscreen joint needs baffles to create a labyrinth. These surfaces block the momentum of the rain, allowing gravity to take over and pull the droplets down and out to the exterior. Such complex joints could be proposed by the aluminum industry because their primary medium is the extruded section. The extrusion of terra cotta allows similar configurations to be made in a simplified form.

From 1979 to 2000 architect Professor Thomas Herzog worked in collaboration with a roof tile specialist, inventor Max Gerha[r]er, to develop a rainscreen façade system using terra cotta tiles.[10] The work was supported by a consortium of German roof tile production plants, given the name Argeton. From 1981 to 1986, Professor Herzog was the development

[9] Architectural Aluminum Manufacturers Association.
[10] Flagge et al., p. 164.

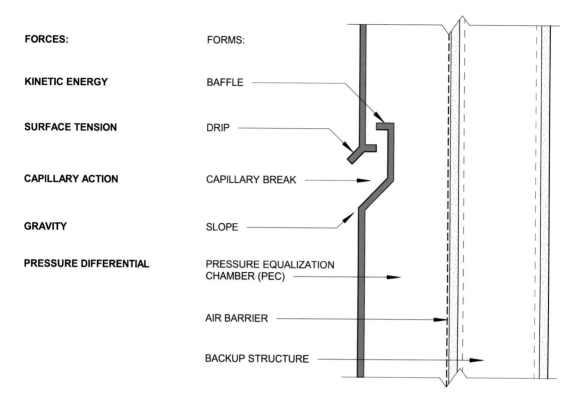

FORCES: FORMS:

KINETIC ENERGY   BAFFLE

SURFACE TENSION   DRIP

CAPILLARY ACTION   CAPILLARY BREAK

GRAVITY   SLOPE

PRESSURE DIFFERENTIAL   PRESSURE EQUALIZATION
                                CHAMBER (PEC)

                                AIR BARRIER

                                BACKUP STRUCTURE

Figure 5.3 Diagram of the "Rainscreen principle for pressure equalized wall design." Specific configurations designed to resist the environmental forces that drive water through cladding systems. Source: Architectural Aluminum Manufacturers Association, 1971.

partner for Moeding, a roof tile plant that brought the first rainscreen tiles to market.[11]

The system was included in an exhibition in Munich of Herzog's works spanning the years from 1978 to 1992, and subsequently published. The tile section is very simple compared to the structural clay units that had been extruded for decades. The distinctive element is a shiplap joint to prevent water entry between horizontal courses of tiles, the edges parallel to the axis of extrusion. The vertical joints are end cuts in this case, and the details indicate that they were blocked by an aluminum spline to catch water entering there and drop it out at the bottom of the wall. The first application, in 1982, was the recladding of an existing market hall for building materials northeast of Munich (Figure 5.4).[12]

The design goals were to create a durable, back-ventilated façade system that could be suspended outside a thick layer of thermal insulation that was, in turn, located outside of the structural frame and infill. The terra cotta units were much lighter in weight than comparably durable alternatives (stone), allowing the mounting rails to be lighter and less expensive. The smaller units were easier to handle during installation and provided

[11] Brookes and
     Meijs, p. 38.
[12] Herzog, p. 47.

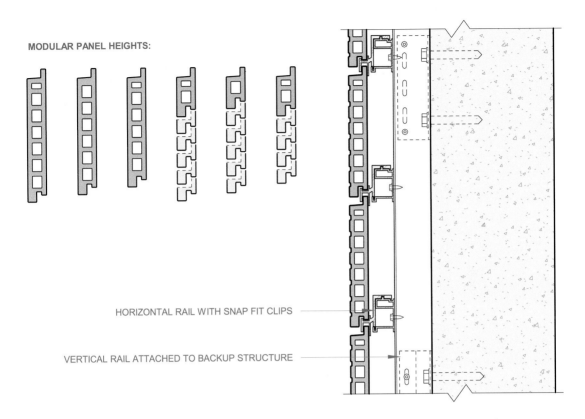

MODULAR PANEL HEIGHTS:

HORIZONTAL RAIL WITH SNAP FIT CLIPS

VERTICAL RAIL ATTACHED TO BACKUP STRUCTURE

Figure 5.4 The terra cotta rainscreen cladding system developed by Thomas Herzog, and others, in the early 1980s. Designed for variable course heights by cutting the modular tiles. Not shown: vertical aluminum splines to block water entry at the ends of the tiles. Source: Herzog.

dimensional flexibility in solving building corners and the insertion of window openings.[13]

This early system demonstrates all of the properties that have become market standards for ventilated terra cotta façades. It further proposed that intermediate coursing heights be achieved by cutting one of three standard shapes such that the rectangular cores replicated the shiplap joints. It was this concept that dictated the simplicity of the terra cotta sections. Today, the industry has grown accustomed to changing the die to control the coursing on a given project. The lasting contributions from Herzog have more to do with the rails and clips than the profile of the clay.[14]

Herzog applied the system to a number of projects during the late 1980s and 1990s, usually in a secondary role. It covered the solid end walls of buildings that were elongated east to west in order to exploit highly glazed and carefully shaded north and south façades. In 2000, the Administrative Tower at the Deutsche Messe, AG, Hannover was published, a landmark building in the history of glass double façades. The solid service cores were clad with a refined version then patented as the Moeding Argeton façade.[15] Horizontal grooves were added on the outer face of the tiles to retard the

[13] Herzog, p. 47.
[14] Brookes and Meijs, p. 39.
[15] Herzog, p. 38.

flow of rain across the surface. They were described as preventing water from being driven upward by higher winds near the top of a taller building. The grooves were also credited with reducing extreme stresses on the tiles during manufacture and adding a strong horizontal emphasis, consistent with aesthetic preferences emerging elsewhere in the field.

Other seminal projects by Herzog include the Design Center, Linz, Austria (1994); housing in Linz (1994–2001); and SOKA-BAU, Weisbaden, Germany (1994–2003). During the same interval, Renzo Piano completed a number of high profile projects that were each developed with a different terra cotta producer. These included Cité Internationale, Lyons, France (1986–1996); housing at Rue de Meaux, Paris (1987–1991); IRCAM, Paris (1988–1990); Banca Popolare di Lodi, Italy (1991–1998); and Potsdamer Platz, Berlin (1992–1999). The notice these projects received from other architects propelled the terra cotta producers to new opportunities.

Complementing the rainscreen principle, building scientists developed another powerful concept for enclosure design during the latter decades of the twentieth century. This concept has been vividly framed and promoted in the writings of Joseph Lstiburek, with full credit to his predecessors. Lstiburek calls it, "The Perfect Wall."[16] Considering the entire hygro-thermal performance, a cladding system begins with rain and air control layers and must add thermal and vapor control. Vapor control is the tough one because conditions vary by climate zone and season. In a cold, dry climate, the vapor drive acts from the warm humid interior toward the cool exterior. In a warm, humid climate, conditioned interior space will be the cool dry side of the wall, and the vapor drive reverses. Many of our preferred settlements are in mixed climate zones where the vapor drive changes direction by season. The "Perfect Wall" works in all climates. The control layers are located together on the outside face of the structure, essentially in the middle of the wall. The permeability of surrounding materials must be chosen to permit the drying of vapor or incidental moisture to either direction as the season permits. Exterior cladding is back ventilated through open joints. While it acts to a variable degree as the water shedding surface, its principal function is to protect the control layers behind it from physical damage and ultra-violet light (Figure 5.5).

The terra cotta wall assemblies promoted by Thomas Herzog and Renzo Piano are ideally suited to the "Perfect Wall." Architects choose the material for the scale and texture that it brings to an opaque surface. Paired with manufacturing economies, this objective produces a range of moderate unit dimensions with relatively frequent connections to a suitable backup wall: structural concrete, concrete masonry infill or engineered steel studs with a durable gypsum sheathing. In each case, there is a robust

[16] Lstiburek.

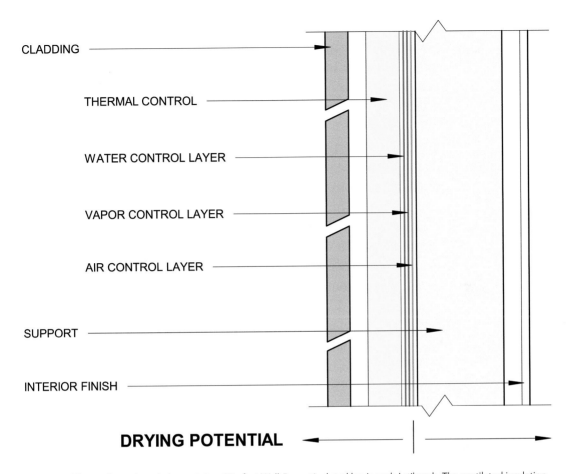

CLADDING

THERMAL CONTROL

WATER CONTROL LAYER

VAPOR CONTROL LAYER

AIR CONTROL LAYER

SUPPORT

INTERIOR FINISH

DRYING POTENTIAL

Figure 5.5 The configuration of elements in a "Perfect Wall," as articulated by Joseph Lstiburek. The ventilated insulation cavity outside the plane of the air/water/vapor barriers provides excellent hygro-thermal performance in all climates. Source: Lstiburek.

exterior plane providing critical support for the water, air and vapor control layers, often combined into a single material. The low water absorption, relatively light weight, and overall durability of terra cotta make it an ideal choice for the free-floating exterior screen. The overlapping girts and rails used to attach the tiles generate the thickness needed for exterior insulation and a ventilated cavity. The insulation material must be suitable for a damp location and the mounting rails and girts are aluminum for corrosion resistance. The only weaknesses in the system are the penetrations of fasteners through the barrier products and the thermal bridge effects due to aluminum sections embedded within the insulating layer. These can be mitigated with careful detailing.

Success in keeping water out of a building is not assured by treating the solid portions of the wall alone. It must include solving the joints, edges and interfaces between systems, particularly where the window meets the wall. Contemporary, standardized terra cotta systems have largely retreated from these difficult locations. The open jointed rainscreen philosophy often assigns the easiest task to the terra cotta, at the center of a wall panel, while the difficult joints at the edges are turned over to other materials. At Rue de Meaux, in Paris, the principal enclosure is carefully articulated glass fiber reinforced concrete panels (GFRC). A terra cotta rainscreen protects the solid portions of the wall, but leaves unprotected the joints between adjacent panels and those between the panel and window systems. The terra cotta is valued for the scale, color and texture it brings to the project, but it does not address the areas where problems are most likely to occur.

Classical terra cotta of the nineteenth century did not retreat from the tough spots. It was asked to take on the roles that carefully cut stones had played, particularly in combination with brick masonry. Stone was used to make a transition from the roughness of masonry to the precision of a window or door. It was used to make exterior sills to collect wind driven rain striking a window and carry it back across the thickness of the wall to the outer surface. One of the consistently troublesome areas in building enclosure is the lower corner of an opening, where the wall, window unit and sill all meet. The solution in stone is a lugged sill. The three-dimensional intersection of the vertical wall and the wash of the sill is resolved in a single block of stone, moving the joints away from the most challenging spot. The geometric freedom of molded terra cotta allowed it to easily and economically accept this role. The plasticity of terra cotta was also used to mitigate problems at simple joints between abutting units. When mortar was the common interface between segments of a sill or parapet coping, the joint was lifted up so that the water is not allowed to sheet across the more absorptive mortar material (Figure 5.6). On copings, the vulnerable mortar joint could be protected with an overlapping cap, called a rolled joint, molded into one of the adjacent units. While very elegant, these joints are no longer recommended out of concern that the overlap will break off. In contemporary work the joints are open, and the mortar has been eliminated. The challenge remains to use both the plasticity and accuracy of terra cotta to shelter critical interfaces and direct bulk water away from them.

Next on Walter Geer's list of terra cotta attributes is lightness. As iron and steel framing developed, the desire to reduce dead loads in the fireproofing of floor span systems was a factor in all markets. Hollow clay tiles offered enormous savings over brick arches between beams. In Chicago, the problem of soil bearing capacity was so significant that the need for lightness extended to the entire structure. In Thomas Leslie's systematic account of Chicago skyscrapers, weight was as great a factor as fire. Buildings were quite literally floated on a layer of hard pan with one hundred feet of water-logged clay below.[17] Foundations consisted of a network of steel beams that spread column loads to concentric portions of the hard pan.

> If exterior walls were re-conceived as systems of piers, they too could be supported on isolated pads, allowing them to be precisely balanced with the lighter iron structures within. The bedrock of New York and Boston allowed large exterior walls to be easily translated into linear foundations, but [architect/builder Frederick] Baumann's isolated pier system resolved these linear elements into point loads, condensing large masonry planes into grids of vertically and horizontally segregated structures that left space for larger windows.[18]

This logic was the first of two stages in the development of the lightweight curtain wall. The second step consisted of material advancements that enabled the "filling in" of iron and steel frames with a veneer, made ever lighter and thinner by the use of glazed terra cotta and larger expanses of glass.[19]

[17] Leslie, p. 7.
[18] Leslie, p. 38.
[19] Leslie, p. 78.

Figure 5.6 Window sill detail at the Carnegie Library of Lincoln, California (1909). Brick and terra cotta produced by nearby Gladding McBean Company. Lugged sills with raised joints designed to resist water entry. Photo: Donald Corner.

The Home Insurance Building (1883), designed in Chicago by William LeBaron Jenny, is generally accepted as the first skyscraper. The less known Tacoma Building (1889) by William Holabird and Martin Roche is an important benchmark with respect to terra cotta. The architects moved the remaining shear walls to the interior, freeing up the street façades. Sanford Loring, President of the Chicago Terra Cotta Company, then proposed tying the terra cotta façade elements directly to metal framing, eliminating the masonry backup wall and saving both weight and floor space.[20] With direct support and anchorage, the terra cotta skin could be as thin as the performance requirements permitted. Exemplary details from the Fisher Building (1896) are included in the case studies of Chapter 9 (see Figure 9.4.7). The ceramic veneers of the 1930s offered a dramatic reduction in weight and cost when compared to traditional, hand packed blocks. Splitting extruded rectangular sections, as described in Chapter 3, produced claddings of 15 pounds per square foot, where the full blocks might have weighed 70 pounds per square foot.[21]

Beyond the unique issues of Chicago, a quest for lighter and thinner cladding systems has characterized all of architecture in the modern era. While this trend favored terra cotta in the pre-modern era, it ultimately contributed to its decline when aluminum curtain wall panels arrived after World War II. Contemporary projects that value the color, texture and scale of clay products still save significant weight by using extruded terra cotta rainscreens rather than anchored brick veneer. Nominally sized hollow core planks by NBK (Terrart Mid) are listed at 10.3 pounds per square foot. Single thickness profiles developed by Palagio Engineering (Terra One) come in at 7.8 pounds per square foot. These contrast with an average value of 40 pounds per square foot for brick veneer.

Exploratory projects in terra cotta have moved performance goals beyond the durable shell that simply resists the elements and into the realm of bioclimatic architecture that mitigates environmental forces acting on the site, the building and the occupants. Since 2016, the University of Buffalo and Boston Valley Terra Cotta have hosted the Architectural Ceramic Assemblies Workshop (ACAW) to probe this frontier. Teams of architects have been invited to conceptualize new terra cotta configurations and take them through to prototypes, with the support of the hosts. Proceedings of the workshops are published annually, describing the investigations in detail.

Terra cotta has three native properties that support bioclimatic performance: porosity, heat capacity, and resilience. The porosity of a ceramic body can be engineered through the clay, the additives and the firing. Low porosity is favored for durability. However, opening up the body to the migration of water and water vapor can return benefits in appropriate climates. The

[20] Leslie, p. 81.
[21] Ferriday, p. 4.

historic reference is an earthenware vessel used to store water, cooled by the evaporation of moisture that has migrated through the shell. Over the years, teams at ACAW have explored evaporative cooling effects from terra cotta screens and paving materials, misted with water in dry climates. They have also posited creating diurnal cycles in humid climates, with excess moisture being absorbed by the clay in the evening and driven out by morning sun.

Due to its specific heat and mass, terra cotta has a significant capacity to collect solar energy. The performance can be improved by extruding volumetric profiles that tilt up toward the sun, optimizing the surface patterns, and applying a dark, matte finish. As return visitors to ACAW, Jason Vollum and teams from AECOM have progressively improved the use of exterior cladding as a solar collector, conveying the heat to storage using high performance tubing embedded in the cores of the extrusion. The heat can be deployed to the interior using terra cotta again as a radiator. The first prototypes used water as the transfer medium, but phase change materials are anticipated (Figure 5.7).[22]

[22] Garófalo and Kahn, 2018, pp. 26–43.

Figure 5.7 Prototype wall module with integrated solar energy collection, exhibited by AECOM at the 2019 Architectural Ceramic Assemblies Workshop. Left: Exterior view with tilted, black ceramic absorbers containing fluid filled tubing. Right: Interior view with ceramic radiators, circulatory pumps and control systems. Photos: Donald Corner.

The resilience of terra cotta in exposed conditions has inspired a wide assortment of shading screens, using tubes, planks or foils. Relatively low weight, good span capability and sufficient strength at points of attachment all recommend the material. Developmental variables include ease of assembly and simplification of the supporting structure. Tension cables and three-dimensional grids offer an efficient means to stabilize the shading elements in space.

A particularly elegant screen wraps the extension of the Holburne Museum, by Eric Parry Architects (2001). The historic main building is a former hotel at the head of Great Pulteney Street, Bath (UK). The extension connects patrons to the pleasure garden behind, with a fully transparent café on the ground level (Figure 5.8). The dappled green façade becomes a solid container of the collection at the top. On the levels in between, light is admitted into the exhibit spaces through a screen wall that connects the terra cotta cladding above and the glazing below with a graceful transition. Ceramic fins, that articulate the cladding modules, drop down to conceal the mullions of the screen. Terra cotta, glass and stainless steel take up their roles as durable components of an exterior assembly (Figure 5.9).[23]

Terra cotta screens have been used to collect rainwater and animate space as the water cascades down the courses. Wall claddings have been designed to contain volumes of soil to support a green façade. At ACAW 2019, Payette Associates displayed complex, backlit wall units designed to trap sound. A team from SHoP Architects exploited the compressive strength of fired clay with interlocking units that make up a barrel vault (Figure 5.10).

As new, environmental uses of terra cotta are proposed, it will be imperative to compare the carbon dioxide equivalent (CO2e) to that of alternative materials. This is a tricky calculation because of the great difference in methods of making the material (kiln types) and the number of high carbon components used to attach it to the building (particularly aluminum). According to the Inventory of Carbon and Energy (ICE) Database,[24] the carbon equivalent of sheet aluminum in Europe, assuming recycled content, is 6.5 KgCO2e/Kg. This is 25.8 times greater than the tabulated value for clay tile (0.255 KgCO2e/Kg). Aluminum is light in weight. Nevertheless, if mid-sized terra cotta panels (10.5 lbs/sqft) are used to replace the 0.125 inch thick (3.17 mm) aluminum face plate on a curtain wall assembly (1.77 lbs/sqft), the material consumption by weight is only 5.9 times greater. This indicates significant savings in carbon equivalent for just the weather face. Adding to this advantage is the expectation of a long service life, and in the right form, the possibility of reclamation and re-use. New assemblies must be developed to reduce the high carbon components associated with terra cotta applications. A flexible, adaptable module suggests a contemporary reinterpretation of the mathematical tiles illustrated at the beginning of this chapter (Figure 5.11).

[23] Heathcote, p. 130.
[24] Jones and Hammond.

Figure 5.8 Gallery extension of the Holburne Museum, Bath (UK) designed by Eric Parry, Architects (2001). Terra cotta and glass screen walls connect the building to its garden setting. Photo: John Rowell.

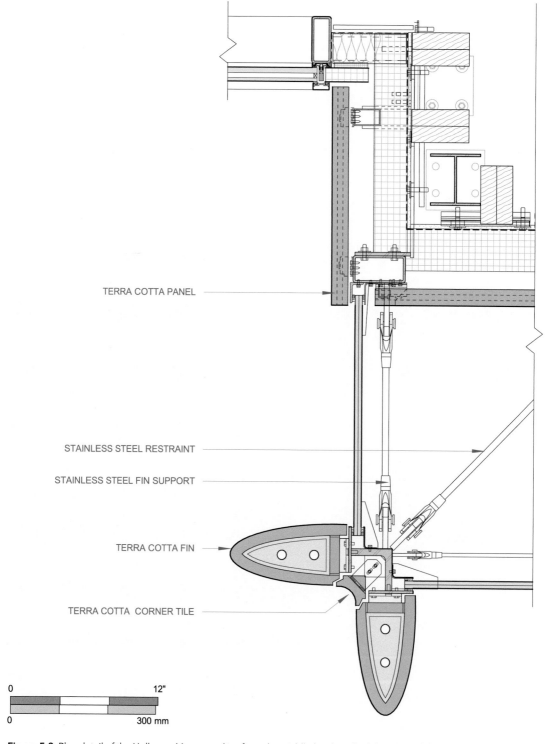

TERRA COTTA PANEL

STAINLESS STEEL RESTRAINT

STAINLESS STEEL FIN SUPPORT

TERRA COTTA FIN

TERRA COTTA CORNER TILE

0   12"

0   300 mm

Figure 5.9 Plan detail of the Holburne Museum taken from the middle level, at the left corner of the façade pictured in Figure 5.8. Vertical terra cotta fins suspended on a stainless steel armature. Timber framed interior walls and finishes not shown. From drawings by Eric Parry, Architects.

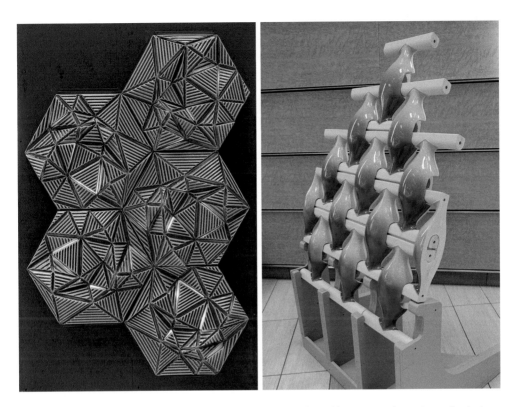

Figure 5.10 Performance prototypes exhibited at ACAW 2019. Left: Backlit, acoustic absorption wall units by Payette Associates. Right: Interlocking parts that form a barrel vault, by SHoP Architects. Photos: Donald Corner.

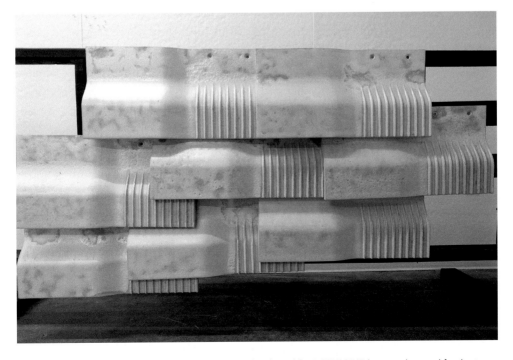

Figure 5.11 "Slide Shingle" wall cladding prototype developed for ACAW 2017 by a student and faculty team from the University of Buffalo and Alfred University. Laura Garófalo, coordinator. Photo: Donald Corner.

CHAPTER 6

Components and Systems

Contemporary applications of terra cotta are dominated by extruded sections applied in a rainscreen assembly, for reasons of production efficiency and long-term performance, as explained in previous chapters. The role of this chapter is to explore, systematically, the many forms that terra cotta components can take and the systems of attachment that are used to integrate these components into logical and effective building enclosures. While extrusion offers efficient production, there is creative opportunity in the other forming methods as well: hand pressing, ram pressing and slip casting. Contemporary application of these techniques will be addressed at the end of the chapter.

The roots of modern claddings are found in the ceramic veneers of the 1920s and 1930s; the so-called "machine made" terra cotta. These roots remain visible at Gladding McBean in California where they were called upon to prototype new claddings for the base of Michael Graves's iconic Portland Building, in Oregon. The City of Portland undertook a complete re-wrap of the largely site cast concrete shell, while striving to maintain the outward appearance of the Postmodern landmark. The base level was originally finished with 9x9 inch ceramic tiles that had broken loose from the backup wall. Michael Graves had originally preferred a terra cotta veneer, inspired by the 1931 Wiltern Theatre in Los Angeles, but it was precluded by the budget[1] (see Figure 4.6).

Gladding McBean was commissioned to prototype a veneer system of 18x18 inch slabs (45x45 cm), mounted in front of a drainage cavity using a generic clip and rail system. One of the design goals was to wrap the piers of the ground floor portico with formed rather than jointed corners. Exploiting the large scale of their extruding equipment, Gladding McBean produced cellular blocks that could be cut apart when finished to provide flat panels with kerfs for mounting. They could also yield two corner panels with short, but monolithic returns (Figure 6.1).

[1] Ceder, 2019.

DOI: 10.4324/9780429057915-7

Figure 6.1 Left: A large-scale extruded block, to be cut apart after firing, producing two 18" [45 cm] square veneer panels, with kerfs for fasteners. Right: Sample of a one piece corner panel cut from a similar extruded block. Gladding McBean Company, Lincoln, California. Photos: Donald Corner.

This prototype offers a glimpse back to the era of the large-scale ceramic veneers, such as those of the McGraw Hill Building in New York. Cross webs of the extruded form were necessary to stabilize the broad surfaces through drying and firing. Traditionally, there was some expectation of distortion in the surface, being in the nature of the material. When the Portland prototype went into full production, warping at the center cell away from the corners was more than the architects could accept, remembering that the previous finish was very flat tiles. The pursuit of flatness, for better or worse, has had much to do with the evolution of terra cotta systems.

RAINSCREEN PANELS

The Banca Popolare di Lodi (1991–8) uses 34 mm (1.34 in) thick, solid tiles that are deeply incised to suggest the refined scale of smaller units. Renzo Piano collaborated with Palagio Engineering of Italy to develop one particular type of open jointed terra cotta screen, expressed as a protective and decorative layer more than a water shedding system.[2] The

[2] Buchanan, 2003, p. 141.

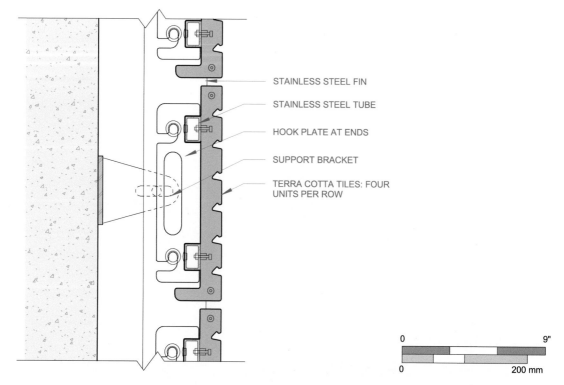

STAINLESS STEEL FIN

STAINLESS STEEL TUBE

HOOK PLATE AT ENDS

SUPPORT BRACKET

TERRA COTTA TILES: FOUR
UNITS PER ROW

0   9"

0   200 mm

Figure 6.2 Vertical section of terra cotta cladding for the Banca Popolare di Lodi, Lodi, Italy (1991–8). Short, extruded segments are back fastened to stainless steel carriers with hook plates at the ends. Vertical stainless steel fins are attached to the primary structure. Designed by Renzo Piano Building Workshop. Drawing based on field notes by the Donald Corner and figures in Buchanan, 2003, p. 143.

screen has a distinct module composed of four tiles, back fastened to a pair of horizontal, stainless steel tubes. The tubes finish with hook plates at each end that drop over pins projecting from vertical steel rails set at 110 cm (3.61 ft) intervals (Figure 6.2). The rows are clustered together in groups of three, with intervening spaces. This produces a horizontal rectangle that is reminiscent in its proportions of the aluminum framed modules of the earlier IRCAM Extension in Paris (1988–9). At IRCAM, and later at Genoa Harbor, the clay units are essentially bricks, pinned to the frame with rods through their cores. Because they are bricks, these other two projects are not illustrated in this book, but Piano's interest in levels of scale through the aggregation of perceived "pieces" is common to all three works (Figure 6.3).

The limited size of the undivided tiles at Lodi served Piano's intent. They were also the largest flat units Palagio could produce at the time. The expressive potential of discrete tiles ganged together with rails and hook plates will be seen at 529 Broadway, New York, by BKSK Architects (see Figure 6.80).

Much later, Renzo Piano undertook the re-use and expansion of a building complex in Milan as headquarters for the business newspaper, *Il Sole*

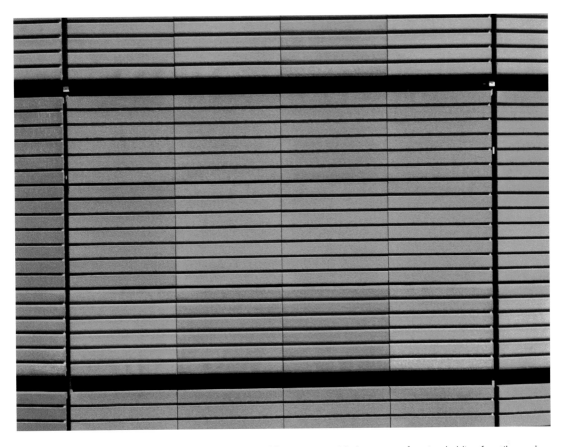

Figure 6.3 Banca Popolare di Lodi. Intermediate scale cladding sections with three rows of carriers holding four tiles each. Larger gaps in the vertical spacing articulate the units. Renzo Piano Building Workshop. Terra cotta by Palagio Engineering, Greve in Chianti (Florence), Italy. Photo: Donald Corner.

24 Ore (1998–2005). By this time, Palagio was working with cellular rather than solid tiles. The complex cross section included deep grooves, similar to those at Lodi, and a horizontal joint between tiles with a male/female interlock that exactly matched the depth of the surface grooves. Limited to 40 cm (1.31 ft) in length, the tiles were butted together in groups of three with expressed vertical joints at 120 cm (3.94 ft). This gave the project the multiple levels of scale characteristic of Piano's work without the complex carrier system of Lodi. The tiles are held by spring clips at the corners for ease of installation, but the short length called for a great number of vertical aluminum rails. With close examination of the façade (Figure 6.4), it is evident that the tiles were not restrained from sliding laterally. The open joints at the module lines come and go as the tiles shift.

The majority of rainscreen systems on the market use simple, shiplap joints that are descendants of Thomas Herzog's prototypes from the 1980s. These tiles were also limited in length, at first, because of the aesthetic preference for a continuous flat surface, free of warping. Competing with

Figure 6.4 Headquarters complex for Il Sole 24 Ore, Milan, Italy (1998–2005). Small-scale extruded tiles are aggregated into intermediate units using vertical gaps and horizontal datum lines. Designed by Renzo Piano Building Workshop. Photo: Donald Corner.

each other, the early producers of rainscreen systems steadily improved their techniques, so that they could extrude and fire longer tiles that would remain flat. Modern plants are highly automated, so the size limits are also controlled by the dimensions chosen for the infrastructure: conveyor belts, drying racks, kilns, etc. At some point, Palagio Engineering chose 1.2 meters (3.94 ft) to be certain that the weight of a plank would not exceed the lifting limit for a single installer working under Italian labor law.[3] Subsequent plants have reached approximately 1.5 meters, or 5 feet, which is a common dimension in façade layouts.

There has also been a continuous evolution of the attachment systems. Early versions derived from fasteners used for stone, but as confidence in the strength of terra cotta grew, smaller, simpler, points of support were developed.[4] A wide range of clips are offered by manufacturers, held in place with hooks, pins, springs, twist locks and screws. The variation reflects increased performance goals tempered by the need to reduce the labor required for installation.

Fastening systems must respond easily to the varying size and weight of the panels. The terra cotta needs to be held securely and prevented from

[3] Palagio Engineering, 2012.
[4] Palagio Engineering, 2019.

rattling. Of particular importance, it must be possible to lift out and re-place an individual tile without disturbing the entire wall. The plan detail (Figure 6.5) shows the vertical joint between two medium sized horizontal rainscreen planks. An extruded aluminum vertical carrier receives clips at the top and bottom of each tile. The continuous vertical gaskets on each side of the joint press the tile against the clips to resist movement. They also work with the carrier to block the entry of water into the cavity. The optional gasket in the joint itself can be used to control the spacing and hide the aluminum carrier if it has not been painted.

Large, heavy panels may require more robust attachments, as in the system offered by Palagio Engineering (Figure 6.6). The vertical carrier is a strong "T" section with cross bars fitted according to the height of the tile. The bracket is designed for anchorage into concrete with the ability to adjust for the desired thickness of the cavity insulation. For design efficiency, only the ultimate clip has to be exchanged, depending on the shape of the tile or the location in the wall (bottom, middle, top). All of the other parts are universal.

It is logical to apply horizontal rainscreen planks as a "stack bond," one above the other. This simplifies the façade layout and reduces the number of vertical rails needed. Four tiles come together at each junction, and with

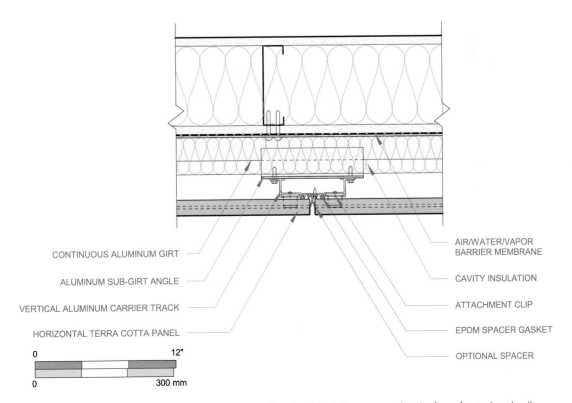

CONTINUOUS ALUMINUM GIRT

ALUMINUM SUB-GIRT ANGLE

VERTICAL ALUMINUM CARRIER TRACK

HORIZONTAL TERRA COTTA PANEL

AIR/WATER/VAPOR BARRIER MEMBRANE

CAVITY INSULATION

ATTACHMENT CLIP

EPDM SPACER GASKET

OPTIONAL SPACER

0   12"

0   300 mm

Figure 6.5 Plan detail showing horizontal rainscreen panels with shiplap joints, mounted to the face of a steel stud wall. Source: NBK North America, Salem, Massachusetts.

Figure 6.6 Showroom sample of the attachment system for heavy, horizontal shiplap panels. Standardized brackets, rails and carrier sections respond to a variety of tile sizes. Palagio Engineering, Greve in Chianti (Florence), Italy. Photo: Donald Corner.

the Palagio system just described, a single cross bar and two clips secure all four corners arriving at that spot.

As it does with brick, stack bond can give the wall a monolithic appearance, rather than the woven character of other bonds. Texture may result from the subtle color variations from one tile to another, particularly with unglazed units. The global texture can be enhanced by exchanging smooth tiles for those with grooves or fine flutes that introduce shadows (Figure 6.7). Within a single bond, brick façades are enriched by blending together several different colors. The same visual opportunities apply to terra cotta, mixing within the full range of body colors, surface treatments and finishes that were introduced in Chapter 4.

The Wisconsin Institute for Discovery (2012), by Ballinger, mixes three surface treatments built on a single, light buff body color. The 12 x 60 inch (30.5 x 152.4 cm) planks may be sandblasted, grooved or wire struck. To specify the blend, the architects developed "terra cotta groups" that are single stacks, fifteen units tall, with varying arrangements of the three finishes. These groups were then applied to the building elevation (Figure 6.8). While a rigorous layout grid is maintained for the elevation, window modules are centered on the terra cotta grid lines, creating stacks of half-length planks that add richness to the overall composition. This does increase the number of vertical mounting rails (Figure 6.9).

The Connecticut College Science Center at New London Hall (2012), by Payette Associates, exploits strong thematic connections to an existing building to enrich the façade composition. New London Hall was constructed in 1914 as the first building on the campus. Realized in a collegiate gothic style, it is clad in rough cut granite from a nearby quarry. The addition is a high performance building that includes careful attention to the sourcing and re-use of building materials. Terra cotta was chosen to meet quality and durability goals, while offering the design flexibility to relate new and old (Figure 6.10).

The existing building has vertical granite masses that are zoned horizontally with contrasting string courses. Distinctive window clusters are combined into multi-story figures on the elevation with spandrels that connect the window modules and proportions. The façade uses simple vertical stacks of rainscreen panels, with three different surface treatments of a warm gray color: sandblasted, machine scored and honed.[5] The south elevation has a projecting volume that houses a staircase. The spandrel zone is reinterpreted with a screen of horizontal baguettes attached directly to the continuous vertical window mullions. The north elevation is a single gable end, with louvers for

[5] Boston Valley Terra Cotta, "Connecticut College."

Figure 6.7 Above: Telling Argeton Rainscreen at Fort Vancouver Regional Library, Vancouver, Washington (2011). Blended, natural colors. Below: Boston Valley Terraclad Rainscreen at the University of Buffalo, Jacobs School of Medicine and Biomedical Sciences, Buffalo, New York (2019). Blended surface textures. Photos: Donald Corner.

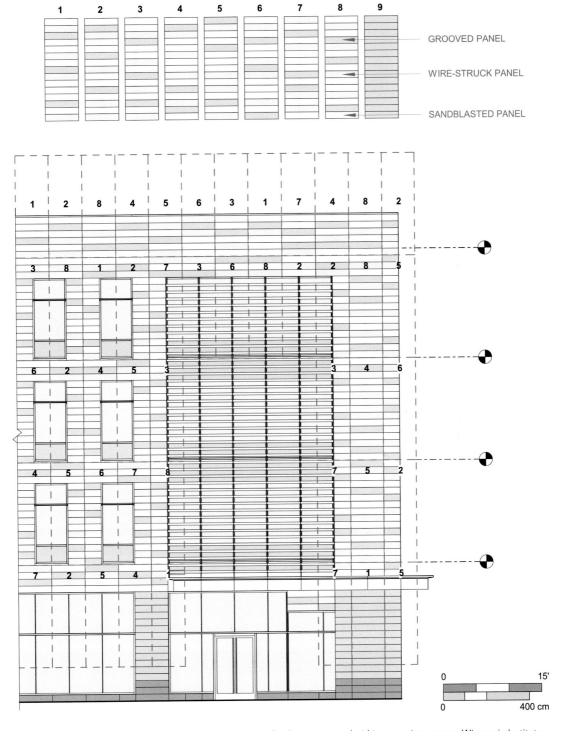

Figure 6.8 Façade composition using three different panel finishes, organized within repeating groups. Wisconsin Institute for Discovery, Madison, Wisconsin. Drawing based on documents by Ballinger, Philadelphia, Pennsylvania.

Figure 6.9 Wisconsin Institute for Discovery, University of Wisconsin, Madison (2012). Sandblasted, wire struck, and grooved Terrart panels supplied by NBK North America, Salem, Massachusetts. Photo: © Zane Williams.

the mechanical systems replacing windows on the upper floor. The spandrel panel at the middle floors is solid but uses a rectangular ribbed profile to pick up the scale and rhythm of the baguette screen. The string courses are created with "half" profiles that, added together, approximate the basic plank module. The larger of the two provides the contrast, while its companion blends in with the adjacent units (Figures 6.11 and 6.12).

The strong influence of datum lines drawn from the original building counteracts the often grid-like appearance of rainscreens. The heights of the terra cotta planks are fine tuned to fit whole courses within each band of the elevation: base, gable, window zone and spandrel zone. This did call for a greater number of extrusion dies, but that price increase has to be considered in proportion to the overall cost and quality of the project. Custom producers encourage architects to modify the course heights of the terra cotta to suit the dimensional requirements of the design, rather than the other way around.[6]

At the M9 Museum in Venice Mestre (2019) there is no hint of a controlling the grid in the terra cotta. The largely opaque walls of the exhibition spaces are wrapped in flowing continuity with a vibrant ceramic

[6] NBK Keramik, 2019.

Figure 6.10 Connecticut College Science Center, New London, Connecticut (2012). A new wing added to the first campus building. Payette Associates, architects. Cladding by Boston Valley Terra Cotta, Orchard Park, New York. Photo: Warren Jagger Photography.

shell. Sauerbruch Hutton Architects accomplished this by deploying slender rainscreen planks in a running bond, featuring a rich mix of colors.

A connecting diagonal path in the site plan of the complex leads to polygonal building volumes with angular cuts through the ceramic skin for access and views. The primary façade of the main building has a long window slot tracing the ascent of a main staircase up to the galleries (Figure 6.13). The tiles are laid out parallel to this opening and the cantilevered volume of the stair (Figure 6.14). The facing administration building also has an inclined expression of its entry. Following the theme, the visually prominent courses of terra cotta are laid out at their modular height, with special cuts occurring only at the ground and the sky (Figure 6.15). The outward looking faces of the complex have a level layout of the openings and cladding, which creates complex cuts where the tiles wrap a corner between the two regimes.

Running bond doubles the density of vertical rails needed to support the tiles. Regardless of slope, the end cuts are always plumb so that they

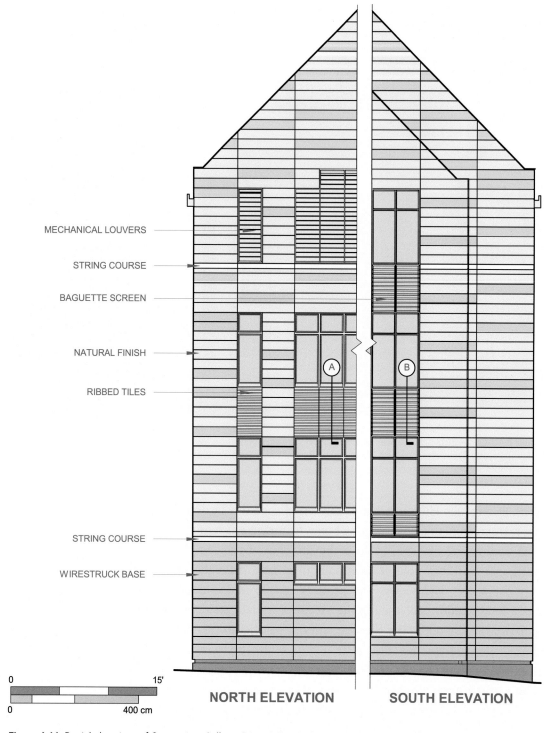

MECHANICAL LOUVERS

STRING COURSE

BAGUETTE SCREEN

NATURAL FINISH

RIBBED TILES

STRING COURSE

WIRESTRUCK BASE

0   15'

0   400 cm

NORTH ELEVATION   **SOUTH ELEVATION**

Figure 6.11 Partial elevations of Connecticut College Science Center. Terra cotta coursing, profiles and finishes are used to relate the new construction to the original New London Hall. The façades also accommodate differences of function: mechanical rooms on the north, an open stair on the south. From drawings by Payette, Associates.

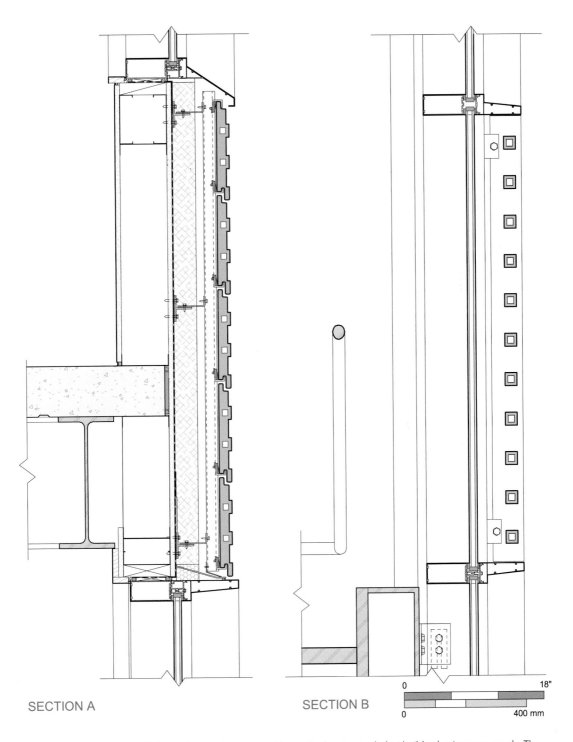

SECTION A   SECTION B

0   18"

0   400 mm

Figure 6.12 At Connecticut College, the spandrel zone on the north elevation is clad with ribbed, rainscreen panels. The south side is a continuously glazed staircase with a baguette screen used to articulate the same façade zone. From drawings by Payette, Associates.

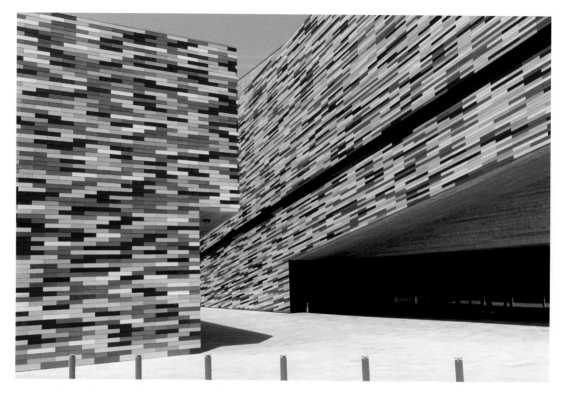

Figure 6.13 Motor entrance to the M9 Museum at Venice Mestre (Venice), Italy (2019). Rainscreen terra cotta applied in a multicolored, running bond. Sauerbruch Hutton, Architects. Terra cotta by NBK Keramik. Photo: Donald Corner.

match the rails. NBK Keramik modified their Terrart Light backing system in response to the light loads and the frequent spacing. Custom design of the clips was required to accommodate the inclined courses. The planks on the larger building are doubled in length, giving them a ratio of 12 to 1 relative to their height. To keep them straight, they were extruded as back to back, twin profiles that were split apart after the first firing.[7]

The colors derive from the historical palette of the Venetian lagoon. Pinks and reds are found in cocciopesto, a traditional lime based plaster that contains ground up bricks and tiles as the large aggregate. A double fired glaze was used, with transparency in the final coat that allows the underlying clay body to show through. The overlay effect adds to the life of the color surfaces that appear directly in front of museum visitors.

Terra cotta rainscreen planks can be applied in a vertical orientation, although horizontal, like courses of bricks, is far more common. Vertical panels can be extruded with overlapping edges to reduce wind driven rain through the tall, open joints. This also blocks the view through to the cavity. The end cuts can be milled to produce overlapping horizontal joints.

[7] NBK Keramik, 2019.

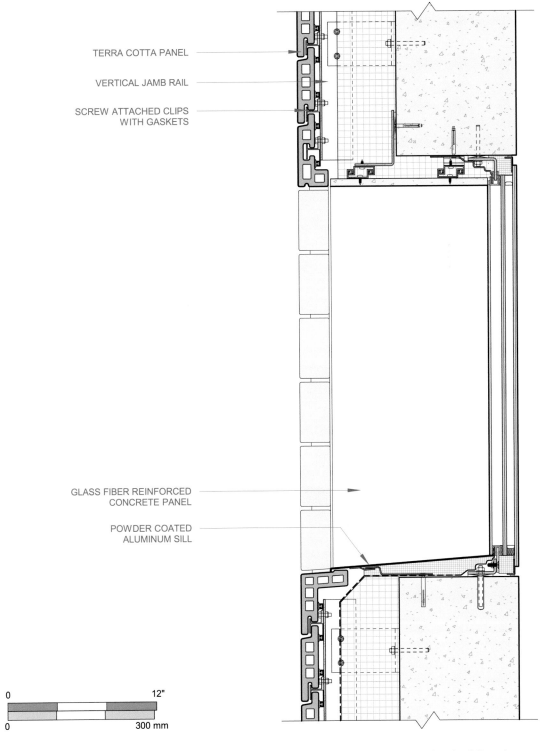

TERRA COTTA PANEL

VERTICAL JAMB RAIL

SCREW ATTACHED CLIPS
WITH GASKETS

GLASS FIBER REINFORCED
CONCRETE PANEL

POWDER COATED
ALUMINUM SILL

0   12"

0   300 mm

Figure 6.14 M9 Museum, Venice Mestre. Custom extrusions used to resolve the recessed slot window that follows the main stair up to the galleries (see Figure 6.13). From drawings by Sauerbruch Hutton, Architects.

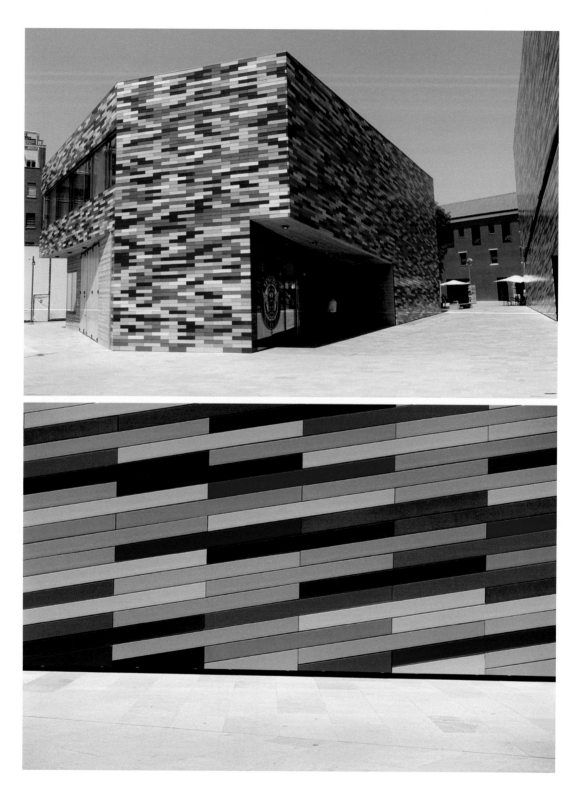

Figure 6.15 M9 Museum, Venice Mestre. Above: Horizontal courses of terra cotta transition to diagonal courses at the corner of the administration building. Below: Custom cut extrusions allow diagonal courses to meet sloping ground at the main building. Sauerbruch Hutton, Architects. Photos: Donald Corner.

The Tykeson Hall case study in Chapter 9 illustrates this detail as it was initially proposed (see Figure 9.5.4). For design reasons, the installed panels have square end cuts, which is the predominant solution. This expresses confidence that the rainscreen layer will keep out most of the moisture, and what does get in will ultimately be dried out by the backside ventilation.

Attachment of vertical panels is accomplished with clips that fit into the cores of the extrusion. Gaskets on the aluminum carriers, or on the clips themselves, prevent rattling in the wind (Figure 6.16). Clips are designed to match the highly standardized wall thickness on the back side of the panel. The same detail applies to deeply profiled vertical panels that will be illustrated later in the chapter.

BAGUETTES

Known also as louvers or foils, baguettes are defined as extruded terra cotta sections that have perfect surfaces on all four sides. There is no back, as in a tile or a plank. Unglazed, there is a dense fireskin on all the surfaces. When glaze is applied, it is also formed equally, all the way around. Paradoxically, these often slender sections can be reliably produced in lengths up to 2 meters (6.56 ft), greater than flat planks that lie against rollers or

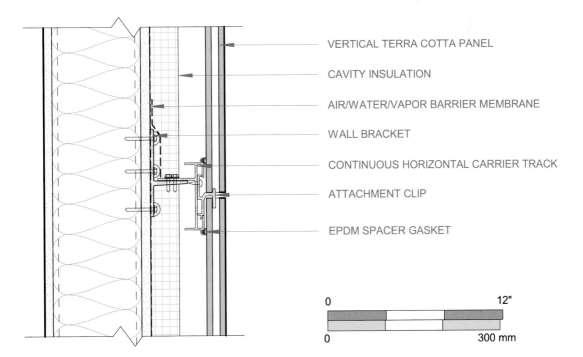

VERTICAL TERRA COTTA PANEL

CAVITY INSULATION

AIR/WATER/VAPOR BARRIER MEMBRANE

WALL BRACKET

CONTINUOUS HORIZONTAL CARRIER TRACK

ATTACHMENT CLIP

EPDM SPACER GASKET

0   12"

0   300 mm

Figure 6.16 Vertical section showing a characteristic attachment of vertically oriented rainscreen panels. Source: NBK North America, Salem, Massachusetts.

racks during production. Manufacturers in this sector have closely held proprietary techniques to bring this about.

There is an "origin" story for the commonly accepted term, baguette. Development work was underway at NBK Keramik on the square shading louvers for Renzo Piano's projects at Potsdamer Platz in Berlin (see Chapter 9). Executive Director Hubertus Fayer carried a yellowish, round tube prototype from Germany to Paris and presented it to Piano, saying, "Monsieur, votre baguette."[8]

Piano had previous experience with the form at the Cité Internationale de Lyon (1985–95). Piano worked with a French manufacturer that eventually became part of the Terreal Group.[9] The basic cladding for the project was a stout, multi-cell rectangular section with a face dimension of 200 mm (7.86 in).[10] It spanned directly to vertical supports at 1.4 meter (4.59 ft) spacing without any of the frames or carriers that were used at IRCAM or Lodi. Secured at the ends, it had more in common with the contemporary baguette than it did the rainscreen planks that were being developed by Thomas Herzog. Piano considered the larger, single module to be appropriate to the scale of the buildings in Lyon. Terreal describes its predecessor firm as having extruded full story height terra cotta elements as early as 1975.[11]

At Lodi, Piano first used spaced, square tube sections as a screen in front of glazed openings (Figure 6.17). These were short lengths, bound together with an internal armature. At Potsdamer Platz, they were made with a single piece (see Figure 9.2.2).

The common form of the baguette is a square section, typically 50 mm or 2 inches on each side. They reduce solar gain at window openings and express a continuity of color and texture when paired with rainscreen panels. They are easily connected though the ends to metal fins at the edges of each façade module (Figure 6.18) Because they are light in weight, they can also be directly integrated with vertical window wall mullions (Figures 6.19 and 6.20).

At the New York Times Building, Renzo Piano and Fox Fowle Architects used a cylindrical shape to reduce the solar gain on the glass façades. The original terra cotta prototypes were eventually replaced with high fire industrial ceramic tubes that have a much higher breaking strength. Since that time, extruded terra cotta strengths have been improved and would likely fill the requirements (Figure 6.21).

Beyond the circle and square, baguettes can be given any cross section that extrusion will allow. Shapes can be designed to selectively reflect or redirect light as the sun angles change. Major manufacturers offer a range of foil shapes for improved shading performance. The extraordinary potential of custom design is demonstrated by James Carpenter's additions

[8] NBK Keramik, 2019.
[9] Lehmann, 2019.
[10] Buchanan, 2000, p. 93.
[11] Terreal North America.

Figure 6.17 Banca Popolare di Lodi, Lodi, Italy. Renzo Piano Building Workshop (1991–8). Baguette screens carry color and texture across window openings on the major façade. Photo: Donald Corner.

to the Israel Museum in Jerusalem (2010). The project merges science and art in creating an experience of light (Figure 6.22).

The Israel Museum is a campus of buildings on an ascending ridge. Moving up, there are views to the landscape and to adjacent gardens, one designed by Isamu Noguchi. Carpenter organized the arrival, visitor services and the ascent with a progression of fully glazed spaces that connect sheltered interiors to the extraordinary, surrounding daylight. Horizontal terra cotta foils prevent direct entry of the sun. The high azimuth angles at 32 degrees latitude allow the north and south façades to be controlled with spaced foils that maintain a view through to the exterior. Counterposed cove and angled surfaces indirectly light the interior (Figure 6.23). On the east and west faces, the foils overlap slightly to completely fill the elevation. Here, the cove and angle work as a projection device. The shapes and colors of trees just outside are re-presented to the visitors on the inside surface of the foils.

Figure 6.18 Lower Sproul Plaza Redevelopment, Berkeley, California (2015). Moore Ruble Yudell, Architects and Planners. Baguette screens are terminated at vertical aluminum fins on the window module. NBK Architectural Terra Cotta. Photo: Donald Corner.

The foils were produced by Moeding Keramikfassaden. The smooth surfaces were protected on their journey through the kilns using a previously fired plank as a refractory support. A white color was chosen to maximize the reflections and the granularity of the surface selected to soften the effects.[12] Aluminum plates are bonded to the ends of the foils to control the alignment against torsion and provide for rapid placement on site; sliding over cleats attached to the vertical mullions (Figure 6.24).

At the Brandhorst Museum in Munich (2009), Sauerbruch Hutton mounted baguettes in a vertical orientation, also for optical effects, in this case related to color. The museum admits daylight into galleries on three levels, including the basement. The ground floor spaces are illuminated with a continuous clerestory window slot that allows the upper story volume to float above the lower. Other apertures are very specific, leaving a

[12] Carpenter.

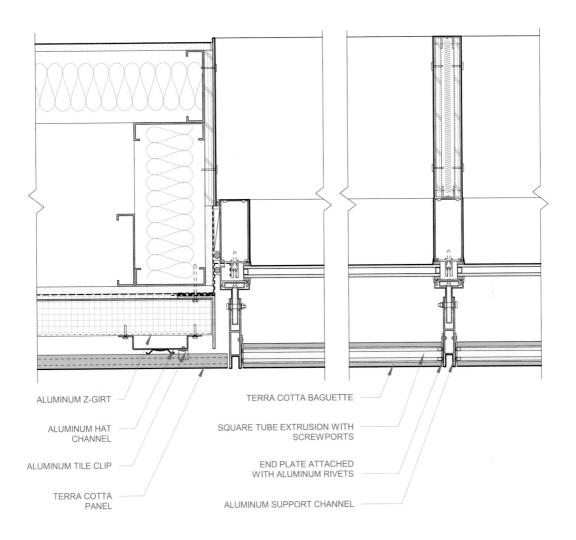

ALUMINUM Z-GIRT

ALUMINUM HAT
CHANNEL

ALUMINUM TILE CLIP

TERRA COTTA
PANEL

TERRA COTTA BAGUETTE

SQUARE TUBE EXTRUSION WITH
SCREWPORTS

END PLATE ATTACHED
WITH ALUMINUM RIVETS

ALUMINUM SUPPORT CHANNEL

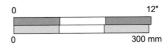

Figure 6.19 Fort Vancouver Regional Library, Vancouver, Washington (2011). Baguettes attach in groups to vertical aluminum supports. Bolted to short lengths of aluminum fin that attach directly through the cover cap on the curtain wall mullion. From drawings by Miller Hull, Architects, Seattle, Washington.

Figure 6.20 Fort Vancouver Regional Library. Miller Hull, Architects. Left: Baguette window screen meets solid horizontal panels. Photo: Donald Corner. Right: Screens meet on support channel bolted to curtain wall. Telling Argeton Rainscreens. Photo: John Rowell.

Figure 6.21 New York Times Building, 620 East Eighth Avenue, New York, NY (2007). Renzo Piano Building Workshop with Fox Fowle Architects. High strength industrial ceramic tubes are used to screen the principal façade planes. Photo: Donald Corner.

Figure 6.22 Israel Museum, Jerusalem (2010). Extruded ceramic foils redirect all of the light on the east and west faces (right). Smaller foils control the north and south faces (center). James Carpenter Design Associates with Efrat-Kowalsky, Architects. Photo: Timothy Hursley courtesy of James Carpenter Design Associates.

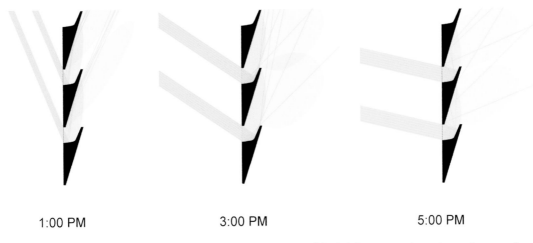

1:00 PM   3:00 PM   5:00 PM

Figure 6.23 Israel Museum. Sun angle studies developing the ability of the foil shapes to project colors and patterns from the exterior on to the inward facing surface of the ceramic. Images: © James Carpenter Design Associates.

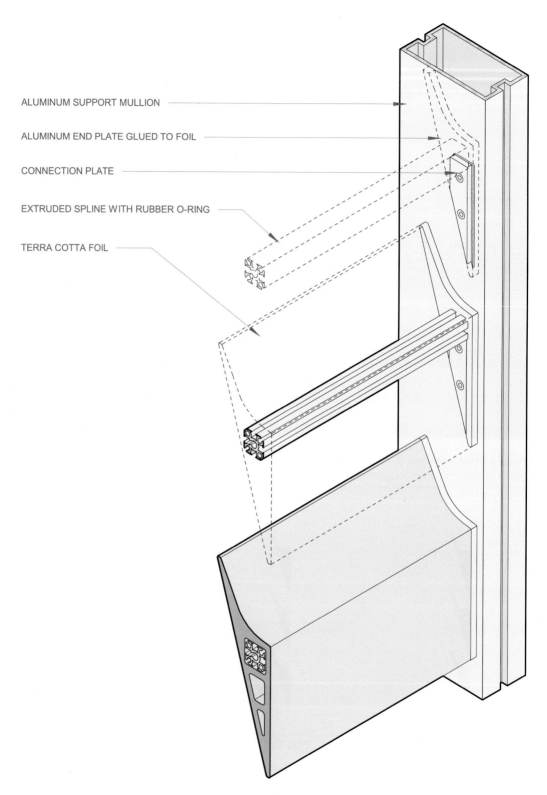

ALUMINUM SUPPORT MULLION

ALUMINUM END PLATE GLUED TO FOIL

CONNECTION PLATE

EXTRUDED SPLINE WITH RUBBER O-RING

TERRA COTTA FOIL

Figure 6.24 Israel Museum. Extruded foils, with central aluminum splines, have shaped plates attached to both ends. These end caps slide over connection plates attached to the aluminum mullions, and they are fixed with a screw. James Carpenter Design Associates. From shop drawings by Moeding Keramikfassaden.

high percentage of solid exterior wall surface to be developed. The terra cotta system creates rich, vibrant façades that provide substance without implying weight (Figure 6.25).

Slender, square baguettes (40 x 40 x 1,100 mm) (1.57 x 1.57 x 43.3 in) are suspended with threaded studs and fasteners through the back surface. They stand in front of a horizontally folded skin of perforated metal, with external insulation and concrete structure behind. The baguettes are glazed with 23 different rich colors, arranged in color groups according to the volumes that they cloak. The metal is finished with alternating bands of color that are seen through the gaps in the

Figure 6.25 Brandhorst Museum, Munich, Germany (2009). Sauerbruch Hutton, Architects. Shaded clerestory window band and fixed vision window within a façade of vertical, multicolored baguettes. NBK Keramik, Emmerich, Germany. Photo: Annette Kisling, Berlin.

baguette layer when viewed straight on. Moving along the façade, as the angle of view changes, the color mixing between the metal and the baguettes changes, embodying what the architects describe as the principle of kinetic polychromy.[13] Viewed obliquely, the baguettes dominate, blending into a smooth realm for each volume. The intention is a façade that acts like an abstract painting, appropriate to the art contained within.

The recessed clerestory window band is controlled on some exposures with either perforated metal shutters or inclined glass shading, operating in the plane of the baguettes (Figure 6.26). Occasional vision windows push through the layers to become flush with the vibrating, dematerialized surface of a given volume (Figure 6.27).

Palagio Engineering offers a rectangular baguette that can be applied either vertically or horizontally (Figure 6.28). It is fastened from the ends using a star shaped extrusion core to resist torsion. This allows the designer to choose the orientation of each piece to the façade plane: parallel, diagonal or perpendicular. The system was applied to the community library at Greve in Chianti, just down the road from the Palagio facility (MDU Architetti, 2001) (Figure 6.29).

The firm Sannini, also in Impruneta, produced short, rectangular sections for a variegated screen that covers the fly tower of the Parco Della Musica di Firenze (ABDR Architetti Associati, 2011). The elements are finished on all four sides and fastened through the ends, like baguettes. The glaze is reminiscent of Granitex as developed by Gladding McBean. As in the earlier era, the savings in production cost and dead weight make this an attractive alternative to an actual stone screen. However, with so many short pieces, the aluminum carriers become a significant part of the visual experience (Figure 6.30).

When using baguettes as shading, architects take delight in the transition of shape or pattern between the open screen and the solid portions of the cladding. The systems-oriented approach is to make the change at a module line where there is an intervening vertical fin. It only remains to reconcile the coursing of the baguettes with the adjacent planks. This was demonstrated at the U.C. Berkeley campus (see Figure 6.18) and the Vancouver Library (see Figure 6.20). The twisted baguettes at the University of Iowa are a dramatic variation (see Figure 3.18).

For the early shading screens at Lodi, Renzo Piano created a zone of transition within a single façade module rather than across the boundary. This was accomplished with a variation on the hook plate carriers that characterize the whole system. Alternating between the straight baguettes there is a returning piece, sliced to match the height of the

[13] Sauerbruch Hutton.

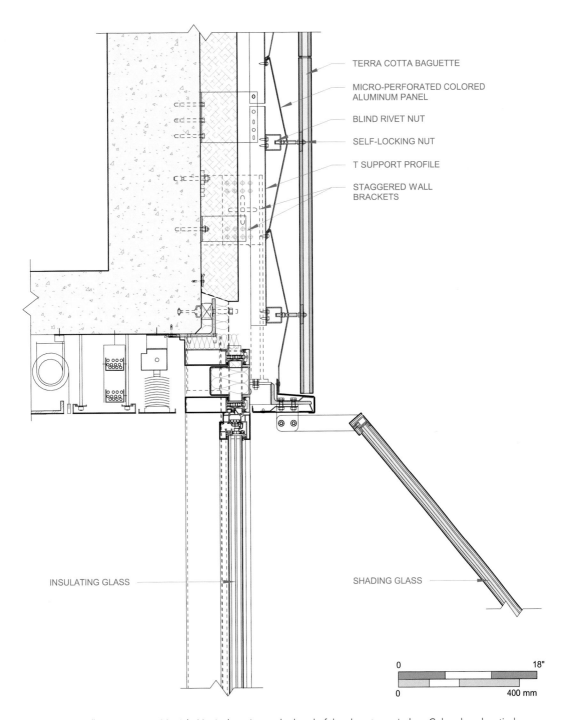

TERRA COTTA BAGUETTE

MICRO-PERFORATED COLORED
ALUMINUM PANEL

BLIND RIVET NUT

SELF-LOCKING NUT

T SUPPORT PROFILE

STAGGERED WALL
BRACKETS

INSULATING GLASS

SHADING GLASS

0   18"

0   400 mm

Figure 6.26 Brandhorst Museum, Munich. Vertical section at the head of the clerestory window. Color glazed vertical baguettes mounted in front of folded aluminum panels, also colored. Sauerbruch Hutton, Architects. From shop drawings by G+H Fassadentechnik, Dresden, Germany.

baguettes. Thus, every other baguette appears to fold inward, across the wall cavity, toward the window. These are held in place by stainless tubes and plates hidden on the back side. It is somewhat a brute force solution

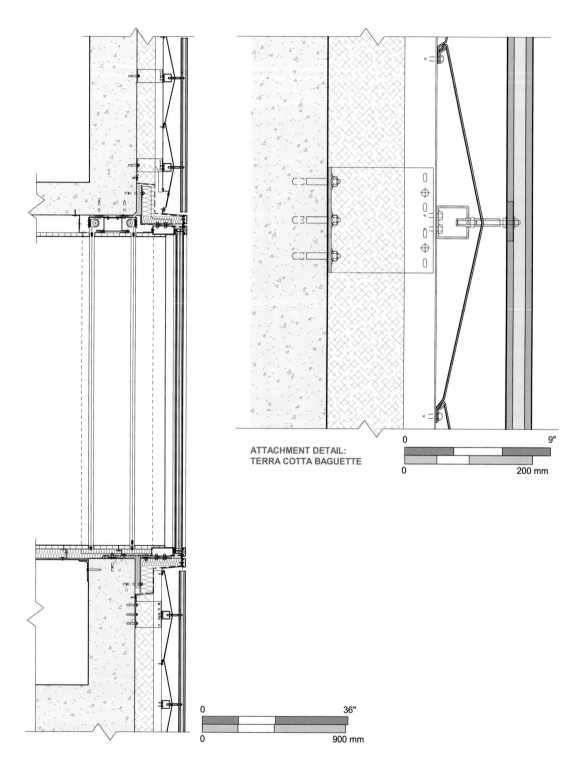

**ATTACHMENT DETAIL:
TERRA COTTA BAGUETTE**

0   9"

0   200 mm

0   36"

0   900 mm

Figure 6.27 Brandhorst Museum, Munich. Vertical section at vision window, flush with the outer surface of the vertical baguettes. Sauerbruch Hutton, Architects. From shop drawings by G+H Fassadentechnik, Dresden, Germany.

Figure 6.28 Extruded baguettes for use as vertical or horizontal shading screens, by Palagio Engineering. Left: Star-shaped core permits a variety of install orientations. Right: Mockup showing flush and diagonal applications. Photos: Donald Corner.

(Figure 6.31). The concept was greatly refined for the subsequent project, Potsdamer Platz, where a custom molded piece of terra cotta receives the ends of the baguettes and also forms the edge of a unitized curtain wall module (see Chapter 9).

Façade composition in the second decade of the current century has been dominated by a particular expression of the non-bearing role of cladding. The construction of panels that span from floor to floor against wind loads is vividly expressed in a full range of materials. Support for the weight of the cladding occurs at each floor, which makes the solid/void pattern of one floor independent of its neighbors. This is expressed on the façades of Discovery Hall at the University of Washington, Bothell Campus (2014) by Hacker Architects (Figure 6.32). In this solution, baguettes are deployed vertically, over a full story, creating a partial screen that makes the transition from closed to open zones on the façade. Budget limitations compelled the use of anchored brick veneer over a substantial percentage of the walls. Terra cotta thus assumes a

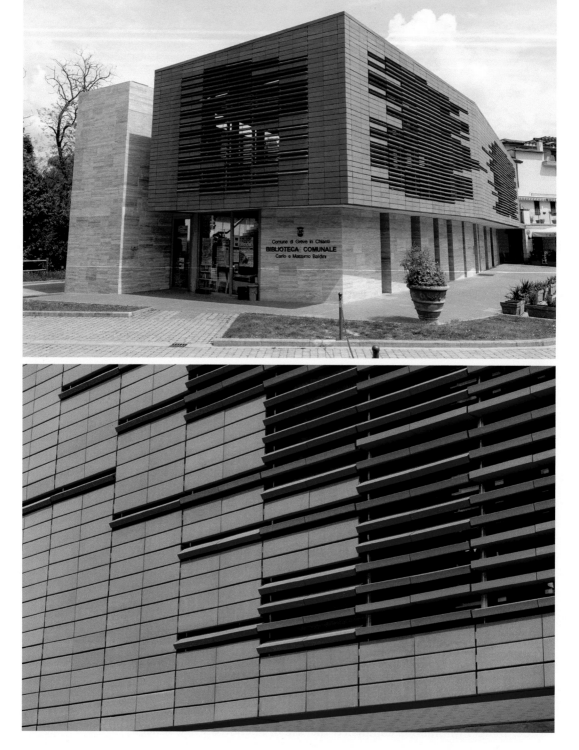

Figure 6.29 City Library, Greve in Chianti (Florence), Italy (2011). Terra cotta screens by Palagio Engineering. MDU Architetti, Greve and Prato, Italy. Below: Detail with baguettes fixed in a diagonal position in front of the windows. Photos: Donald Corner.

Figure 6.30 Parco Della Musica di Firenze, Florence, Italy (2011). Designed by ABDR Architetti Associati. Above: Ceramic screen wraps the fly tower as seen from the roof top performance space. Below: Detail of the elements, with mottled glaze. Terra cotta by Sannini, Impruneta (Florence), Italy. Photos: Donald Corner.

Figure 6.31 Banca Popolare di Lodi, Lodi, Italy (1991–8). Designed by Renzo Piano Building Workshop. Left: Mockup of the stainless steel carrier system with alternate baguettes returning toward the window. Right: Completed view on site. Pattern of baguettes coordinated with profiles on adjacent panels. Palagio Engineering. Photos: Donald Corner.

traditional role in this context, that of a refined mediator between the brick and the glass (Figure 6.33).

EXTRUDED PROFILES

Thomas Herzog's flush rainscreen panels were first used in highly rationalized systems, often to clad the solid east and west ends of elongated buildings that had active glass façades on the north and south, as at SOKA-BAU (Weisbaden, Germany, 1994).[14] Renzo Piano articulated the volumes of his buildings using rounded corners and returns at the windows to bring depth to the façade. Under the economic pressure of commercial development, contemporary terra cotta façades do not always get elaborated in the details to relieve the flat and thin appearance that can result from a literal, off-the-shelf rainscreen. While preserving the logic of the rainscreen plank, architects have turned more and more to bold, custom profiles to add shadow and depth to the façade. For the town center at Scandicci

[14] Flagge et al., p. 141

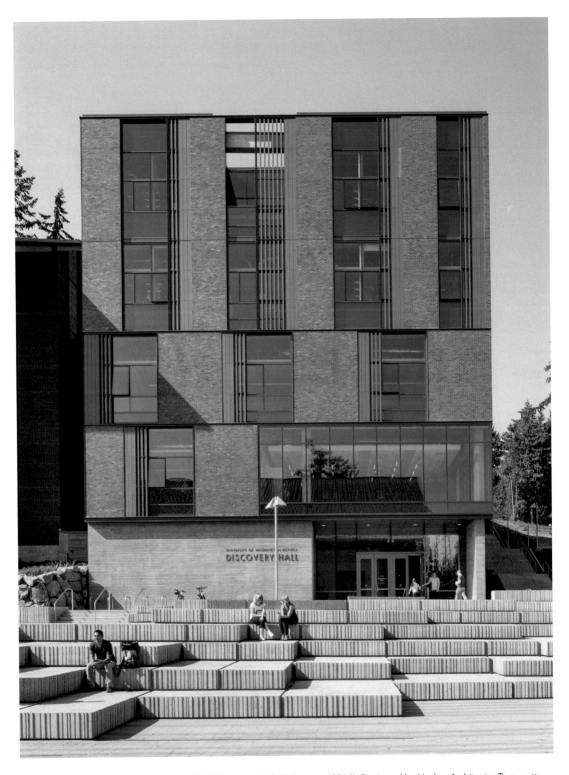

Figure 6.32 Discovery Hall, University of Washington, Bothell Campus (2014). Designed by Hacker Architects. Terra cotta by Telling Architectural. Photo: © Lara Swimmer.

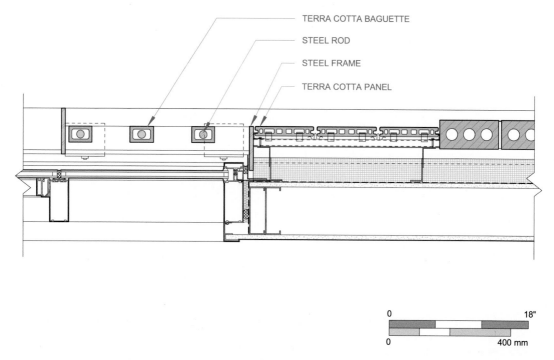

TERRA COTTA BAGUETTE

STEEL ROD

STEEL FRAME

TERRA COTTA PANEL

0   18"

0   400 mm

Figure 6.33 Discovery Hall, UW Bothell. Plan detail of cladding transition: anchored brick veneer, vertical rainscreen, vertical baguette screen, operable vision window. From drawings by Hacker Architects, Portland, Oregon.

(Florence, Italy, 2013), Rogers Stirk Harbour developed strong shadow lines with a profile that is still an explicit rainscreen, shedding the water outward with no ledges. This is one example of a whole class of solutions for which the fasteners fit into the back of the tile, rather than working off the edges (Figure 6.34).

Extrusions can also be given a full bodied, ornamental character, as they were at 10 Bond Street, by Selldorf Architects (2016). Located just north of Houston Street in Manhattan, the building responds to the rich materiality and depth of the façades in the historic neighborhood (Figure 6.35). Spandrel panels on the rectilinear façade are clad with two courses of ogee curves. One is applied upright (cyma recta) and below it the same shape is inverted (cyma reversa) (Figure 6.36). The columns are clad with a concave profile and some flat, return panels. Dramatic volumetric effects are achieved with only two major profiles. Boston Valley Terra Cotta developed a rich "russet" glaze that complements the weathering steel used at the commercial base, trellised top and window surrounds (Figure 6.37).

Bold profiles are more often applied vertically, like folds in a curtain. This reflects the internal logic of an enclosure wall, as it spans from floor to floor. It also provides for a simple and secure means of attachment. The flat back of the profile acts as a shear panel to pick up all the loads of the

Figure 6.34 Town Center, Scandicci (Florence), Italy (2012). Rogers Stirk Harbour, Architects. Left: Mockup of profiled panels at Palagio Engineering. Right: Texture and shadow on completed elevation. Photos: Donald Corner.

Figure 6.35 10 Bond Street, New York, NY (2016). Designed by Selldorf Architects. Ceramic cladding by Boston Valley Terra Cotta. Photo: Nicholas Venezia courtesy of Selldorf Architects.

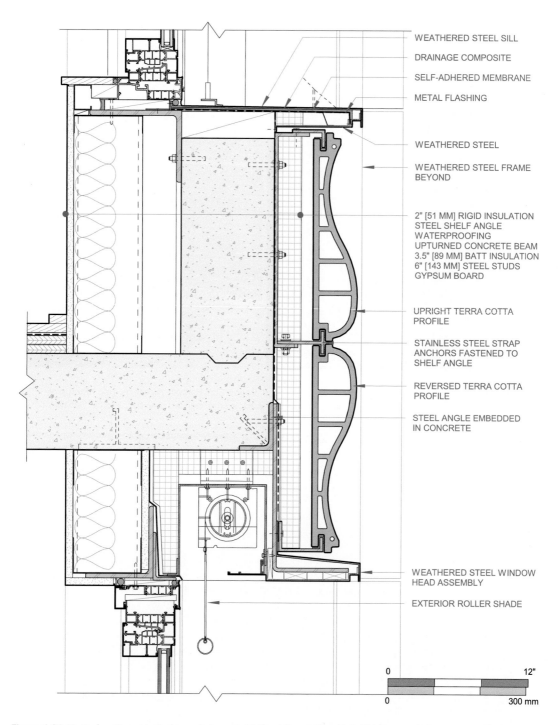

WEATHERED STEEL SILL

DRAINAGE COMPOSITE

SELF-ADHERED MEMBRANE

METAL FLASHING

WEATHERED STEEL

WEATHERED STEEL FRAME
BEYOND

2" [51 MM] RIGID INSULATION
STEEL SHELF ANGLE
WATERPROOFING
UPTURNED CONCRETE BEAM
3.5" [89 MM] BATT INSULATION
6" [143 MM] STEEL STUDS
GYPSUM BOARD

UPRIGHT TERRA COTTA
PROFILE

STAINLESS STEEL STRAP
ANCHORS FASTENED TO
SHELF ANGLE

REVERSED TERRA COTTA
PROFILE

STEEL ANGLE EMBEDDED
IN CONCRETE

WEATHERED STEEL WINDOW
HEAD ASSEMBLY

EXTERIOR ROLLER SHADE

| 0 | | 12" |
| 0 | | 300 mm |

Figure 6.36 Vertical section at typical spandrel panel. 10 Bond Street, New York, NY. Reversed terra cotta profiles integrated with weathered steel sills and flashing. From drawings by Selldorf Architects.

projecting shapes. Clips reach into the end cuts of the extrusions where the material is predictably 10 mm (.394 in) in thickness. If larger clips, with more bearing area, are needed, the back wall of the extrusion can be given closely spaced ribs inside the cavities to match the dimension required.

Figure 6.37 10 Bond Street. Detail of spandrel and column covers. Russet glaze by Boston Valley Terra Cotta, harmonizes with weathered steel sills and flashings. Photo: © Christopher Payne/Esto.

The Nano Institute, in Munich (2019), by kleyer.koblitz.letzel.freivogle. architekten, is a high performance laboratory building on a sensitive site, adjacent to the revered English Garden (Figure 6.38). The upper façades are restrained, monochromatic bands of terra cotta between recessed window strips that provide views from a variety of workplaces (Figure 6.39). There is a darker glaze at the base. The building is given depth and texture with randomized vertical profiles inspired by the trapezoidal corrugations found in metal siding. These basic shapes were elaborated into two profiles that could be combined to produce effective levels of contrast and variation. Applied in pairs, these shapes produce four combinations by inverting the pieces end for end. The pairs themselves could then be inverted to yield four more combinations (Figure 6.40).

The building requires significant conditioning, with air change needed at the mechanical bay on one side of the upper floor. Industrial metal louvers would not have been appreciated in the neighborhood, so they are replaced with a screen of rhombic baguettes that take over the window band at that level. The flat surfaces and receding angles of

Figure 6.38 Nano Institute, Munich, Germany (2019), Designed by kleyer.koblitz.letzel.freivogle.architekten. Vertically oriented, profiled terra cotta, randomized to produce texture. NBK Keramik, Emmerich, Germany. Photo: NBK.

the baguettes harmonize with the shapes generated in the other profiles (Figure 6.41).

The Alaska State Library, Archives and Museum (2016), by Hacker Architects, has a contoured surface that is articulated by the joints rather than the profiles. The convex and concave folds in the terra cotta are aligned from course to course rather than randomized. Instead, the location of vertical joints is staggered in a running bond, extending as far as each unitized panel would allow (Figure 6.42).

Located in Denver, Colorado, the Kirkland Museum of Fine and Decorative Art (2018) was designed by the Seattle firm of Olson Kundig. The central mass of the building is clad with courses of vertically oriented profiles to produce a seemingly random texture, in this case energized by the use of brilliant yellow colors. The color adds prominence to a small building that stands near a group of much larger, landmark museums. The exterior reflects the collection within, a mix of art and craft items that are often textured and colorful (Figure 6.43).

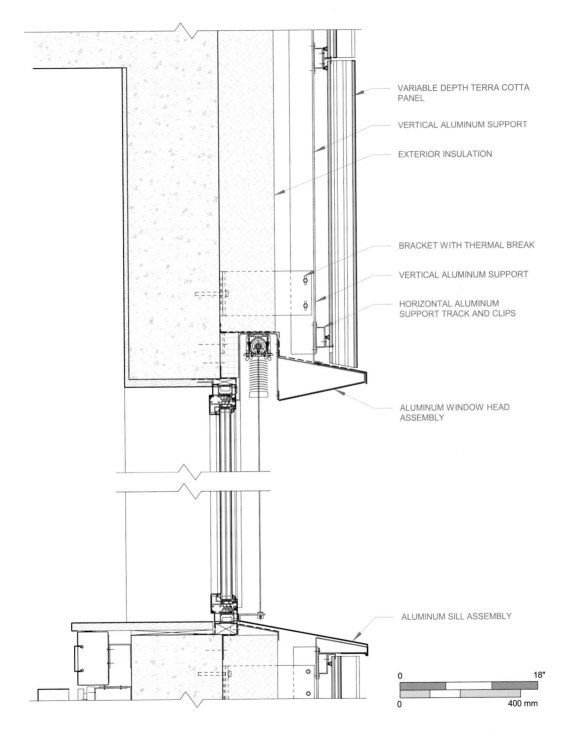

VARIABLE DEPTH TERRA COTTA PANEL

VERTICAL ALUMINUM SUPPORT

EXTERIOR INSULATION

BRACKET WITH THERMAL BREAK

VERTICAL ALUMINUM SUPPORT

HORIZONTAL ALUMINUM SUPPORT TRACK AND CLIPS

ALUMINUM WINDOW HEAD ASSEMBLY

ALUMINUM SILL ASSEMBLY

0   18"

0   400 mm

Figure 6.39 Nano Institute. Vertical section at recessed horizontal window bands with aluminum head and sill transitions, and variable depth rainscreen profiles. From drawings by kleyer.koblitz.letzel.freivogle.architekten.

Architect Jim Olson derived the texture from a concurrent project in the Northwest for which he clad the walls with rectangular profiles in wood that vary in dimension and orientation. In Denver, the concept

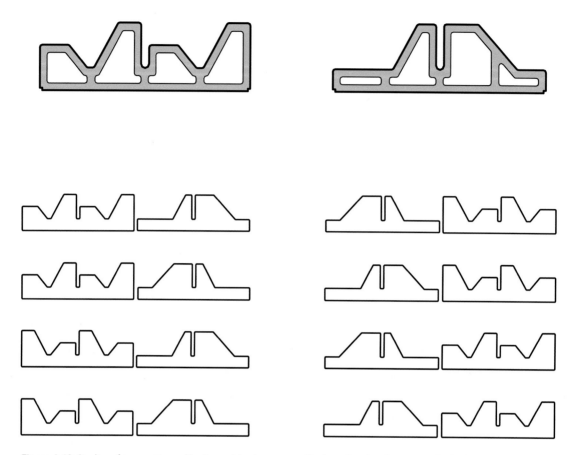

Figure 6.40 Studies of terra cotta profiles in combinations created by inverting the elements, individually and in pairs, to produce a randomized texture. From drawings by kleyer.koblitz.letzel.freivogle.architekten.

was translated to terra cotta extrusions, an appropriately decorative human product capable of sustaining strong color.[15] The system consists of closely spaced baguettes so that it was possible to independently exchange both the shapes and the colors. The wall is laid out in horizontal bands at 2'-4-1/2" (72.39 cm) in height. The baguettes are 4'-9" (144.78 cm) tall, so that they stitch the bands together. It is the vertical equivalent of the running bond at M9, Mestre (see Figure 6.13). The patterns are carefully composed and do eventually repeat. There are four different groupings, each 4'-9" (144.78 cm) in width, making the repetition impossible to detect.

Adding to the sparkle, the baguettes are occasionally replaced with bars of solid art glass that have gold metallic paint on the back face. These are fitted with metal bezels at the ends that receive the same mounting clips used for the terra cotta. The client specified windowless gallery space, as a jewel box. There are display vitrines integrated into the façade pattern to give passersby a hint of the collection (Figure 6.44).

[15] Olson.

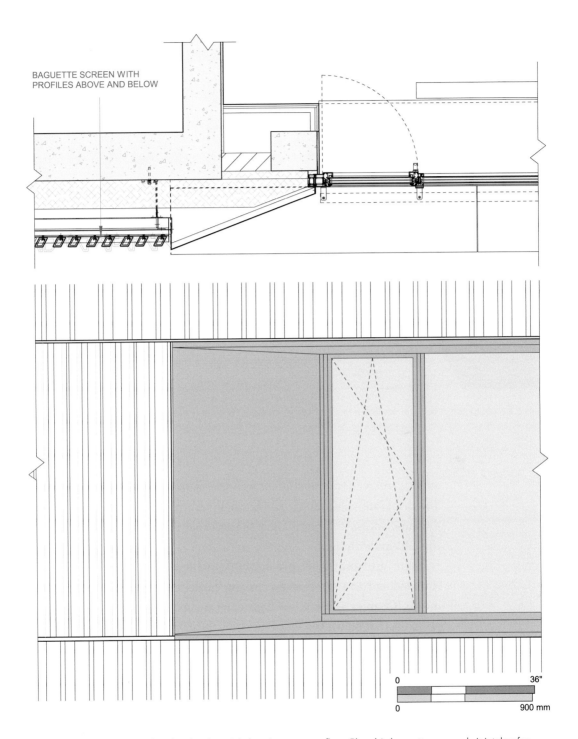

BAGUETTE SCREEN WITH
PROFILES ABOVE AND BELOW

0   36"
0   900 mm

Figure 6.41 Nano Institute. Plan detail and partial elevation at upper floor. Rhombic baguettes conceal air intakes for building mechanicals. From drawings by kleyer.koblitz.letzel.freivogle.architekten.

The Roberts Pavilion (2016) at Claremont McKenna College, in Southern California, uses vertically oriented, extruded profiles to clad large-scale walls that curve and fold in plan. The Pavilion is a collection of

Figure 6.42 Alaska State Library, Archives and Museum (2016). Designed by Hacker Architects. Unitized curtain wall using folded terra cotta profiles with staggered joints. NBK North America. Photo: Hacker Architects, ECI Hyer.

athletic, recreation and student wellness spaces offered as a social center. The disparate volumes are unified behind curved surfaces that envelope dynamic access and circulation spaces. Bold projections and cut-outs fill the interior with light.

There are numerous projects that achieve long radius curves and rounded corners by post-processing horizontal planks by proprietary means (Chapter 3). Here, the flush exterior surface is composed of vertical elements that retain the shape given by the extrusion die (Figure 6.45).

John Friedman and Alice Kimm, architects (JFAK), worked with NBK North America to develop the approach. The basic cladding is a mix of Terrart Large, rainscreen panels in 10, 20 and 30 inch widths (25.5, 50.8, 76.2 cm). They are arranged along courses that are six feet tall (1.83 m). The stone colored glaze is occasionally exchanged for red and white, which activates the façades with the school colors.

Within this vocabulary, the architects developed inside radius and outside radius curves through 45 and 90 degrees. There were practical limits on the overall thickness of the profile, measured from its flat, back side. This translated into minimum allowable radii, and numbers of repeating

Figure 6.43 Kirkland Museum of Fine and Decorative Arts, Denver, Colorado (2018). Olson Kundig Architects. Vertical baguettes with different sizes and orientations. NBK North America. Below: Detail with display vitrines and art glass inserts within the terra cotta. Photos: Alex Fradkin.

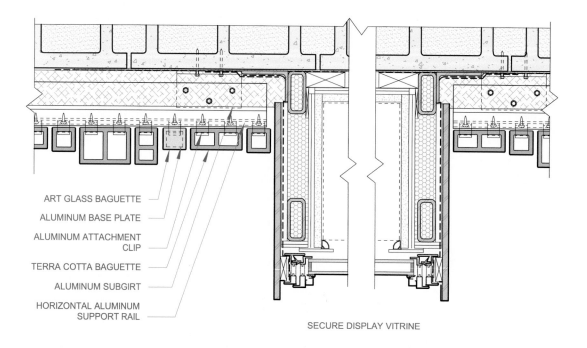

ART GLASS BAGUETTE

ALUMINUM BASE PLATE

ALUMINUM ATTACHMENT
CLIP

TERRA COTTA BAGUETTE

ALUMINUM SUBGIRT

HORIZONTAL ALUMINUM
SUPPORT RAIL

SECURE DISPLAY VITRINE

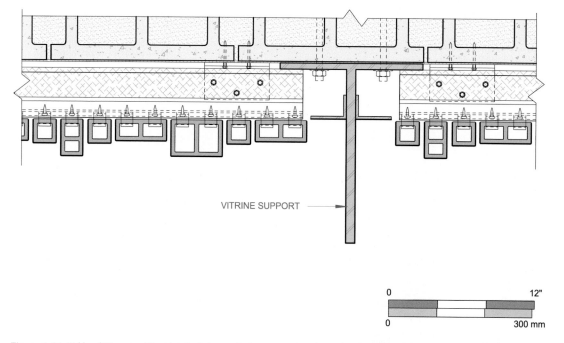

VITRINE SUPPORT

0		12"
0		300 mm

Figure 6.44 Kirkland Museum. Plan detail of the vertical terra cotta and art glass baguettes applied to a CMU wall. Construction shown for the display vitrine and the visible support below it. From drawings by Olson Kundig Architects, Seattle, Washington.

elements that would be needed to accomplish a bend. The 90 degree bend over three panels is the most demanding (Figure 6.46). As the design developed, the architects were able to adjust the plan to reduce the number

Figure 6.45 Roberts Pavilion at Claremont McKenna College, Claremont, California (2016). Jonathan Friedman and Alice Kim (JFAK), architects. Large, vertically oriented panels that form a lightweight, undulating façade, suspended over large areas of glazing at the lower levels. Photo: Benny Chan, Fotoworks.

of different radii required, therefore decreasing the number of custom shapes.[16]

St. Albans Place in Leeds, UK, uses large profiled panels in both horizontal and vertical orientations. Feilden Clegg Bradley Studios took over a repeatedly delayed housing project for the client, VITA Student. The site called for a long, rectangular plan to enclose park space and screen out noise from an adjacent motorway. Residential density goals called for three attached towers, with the center section reaching 18 stories. There was a need to articulate the elevations to mitigate the proposed jump in scale at the edge of the city center. Leeds is known for the production and local use of ceramic tiles and faience (glazed terra cotta). It is also known for textiles. These two industrial craft traditions inspired the design narrative for the façades. Vertical and horizontal panel orientations interlock as the "warp" and "weft" of a woven fabric (Figure 6.47). The specific pattern of the weave changes across the three parts of the building, as does the color of the glaze. Contrasting strands are created within each fabric element

[16] Friedman.

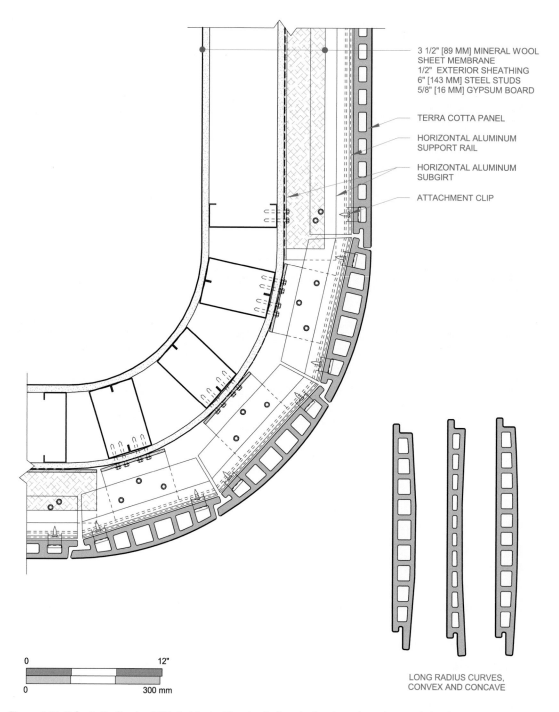

3 1/2" [89 MM] MINERAL WOOL
SHEET MEMBRANE
1/2" EXTERIOR SHEATHING
6" [143 MM] STEEL STUDS
5/8" [16 MM] GYPSUM BOARD

TERRA COTTA PANEL

HORIZONTAL ALUMINUM
SUPPORT RAIL

HORIZONTAL ALUMINUM
SUBGIRT

ATTACHMENT CLIP

LONG RADIUS CURVES,
CONVEX AND CONCAVE

0   12"

0   300 mm

Figure 6.46 Roberts Pavilion by JFAK, Architects. Plan detail of vertically oriented panels extruded with convex or concave faces to make up long radius curves on the building façade. From shop drawings by NBK North America.

by changing the intensity and orientation of the terra cotta profiles. Differences of color and shadow are enhanced by pooling effects in the glaze.

The building is constructed with concrete floor slabs, infilled at the perimeter with a steel framing system. The cement particle board sheathing

Figure 6.47 St. Albans Place, Leeds, West Yorkshire, UK (2020). Feilden Clegg Bradley Studios, architects. Student housing complex with a "woven" pattern of profiled terra cotta, combining horizontal and vertical orientations. Pooled glaze effects by NBK Keramik. Photo: Richard Battye/FCB Studios.

allows vertical aluminum rails to be laid out as needed, without being tied to the pattern of wall studs. Mounting clips for the rails reach through an extra layer of cavity insulation required for sound attenuation. Horizontal hat channels of varying sizes receive the clips that hold the terra cotta. The woven pattern of the tiles adds complexity to the backing system, particularly to provide for deflection gaps that work their way across each floor line. The vertical tiles are 600 mm (1.97 ft) and 1800 mm (5.9 ft) tall. They are deep enough to span these distances under wind load, but need additional support for impact loads, such as those that might be imparted by a maintenance scaffold. To meet these requirements there are additional rails and channels at midspan, faced with cushions that push back against external loads. Taken altogether, these framing installations added a significant labor component to the cost of the cladding (Figure 6.48).

The architects specified single fired terra cotta to reduce the cost and the carbon footprint. The thickened edges of each profile and the weaving pattern prevent the unglazed end cuts of the terra cotta from being seen. Cuts just above the ground floor glazing are covered with a matching, profiled plate. NBK was able to formulate glazes with color pooling under single

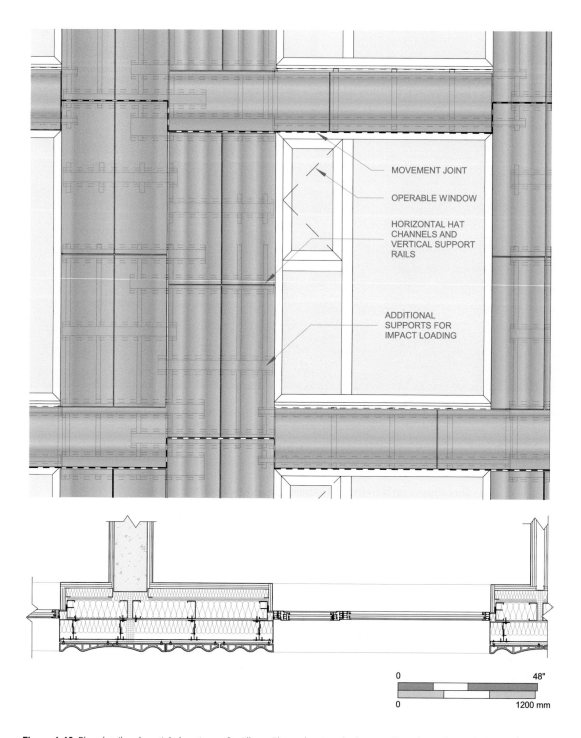

MOVEMENT JOINT

OPERABLE WINDOW

HORIZONTAL HAT
CHANNELS AND
VERTICAL SUPPORT
RAILS

ADDITIONAL
SUPPORTS FOR
IMPACT LOADING

| 0 | 48" |
| 0 | 1200 mm |

Figure 6.48 Plan detail and partial elevation at St. Albans Place, showing the layout of hat channels needed to support the "woven" façade. FCB Studios, architects. From shop drawings by FK Group, Altrincham, Greater Manchester, UK.

fire. The lighter tones exposed on the profile ridges of the building wings match the dominant color of the central tower.[17]

CONSTRUCTION STRATEGIES

There are four basic strategies for the application of terra cotta to a building façade: hand set, panelized on studs, unitized curtain walls and precast concrete backed systems. Hand setting of the terra cotta is as the term implies. The subcontractor is generally responsible for everything outside of the backup wall and barriers, with the backup wall being structural concrete, concrete masonry, or most often steel studs and sheathing. The cost of steel studs infilling a structural frame should be added to the estimate when comparing hand set to other systems. The studs may be thicker than normal construction, up to 16 gauge, or stronger. Girts, rails and clips are attached to the backup wall and workers, on scaffolds or lifts, install the terra cotta panels into the clips.

The cost of hand set work varies a great deal by region. With more projects and more qualified subcontractors, prices on the East Coast of the United States can be 25–30% lower than on the West Coast.[18] The complexity of the design can be a significant factor if it results in more complex systems of attachment. The volume of the project, height of the building, availability of space for staging and the participation of union labor are all major variables.

In Massachusetts, the Boston architectural firm of Bruner/Cott Associates designed 45 Province Street (2009) using a terra cotta rainscreen applied offsite to panelized steel stud walls. The building is 31 stories tall on a tight urban site in the Mid-town Cultural District. Red terra cotta was chosen early in schematic design to harmonize with the context. In Boston, brick would have cost less as a skin, but the structural costs would have been increased because of the weight. Terra cotta offered more freedom to step back the façades as the building rises, because the lighter loads could be supported on the cantilevered edges of the floor slab, without spandrel beams or strict adherence to the column grid. In general terms, the cost of panelized systems begins at the upper end of hand set systems, but there are compensatory savings in project delivery time.

The panelized stud walls operate with the same principles as a unitized curtain wall. Gravity and wind loads are transferred from the top of the

[17] Richardson.
[18] Streff.

panel to the floor slab. Wind loads at the bottom of the panel are pinned to the panel below (Figure 6.49). These panels were craned into place from the street at night and plumbed up in the daylight. Multi-layered sealant joints were applied from suspended, window washing scaffolds. The contractor used different colors of sealant in each layer to assist the quality control consultant in verifying the work. Because of the movements anticipated at the horizontal joints, the interior finishes could not be attached directly to the back of the prefabricated panels. An interior layer of studs was added.

The terra cotta covered walls are expressed as a slight projection where the façade steps back in plan to create residential balconies (Figure 6.50). To accomplish this, the internal studs are decreased in depth, allowing continuous insulation of all the exposed surfaces. The perpendicular glazing unit is aligned with the edge of the cantilever slab to facilitate a vertically continuous waterproof joint. This detail was selected by performance testing alternative mockups (Figure 6.51).[19]

Terra cotta and unitized curtain wall systems are logical partners, particularly for large-scale projects on congested urban sites. Terra cotta offers a vastly greater range of durable colors, forms and textures than do

[19] Crosby.

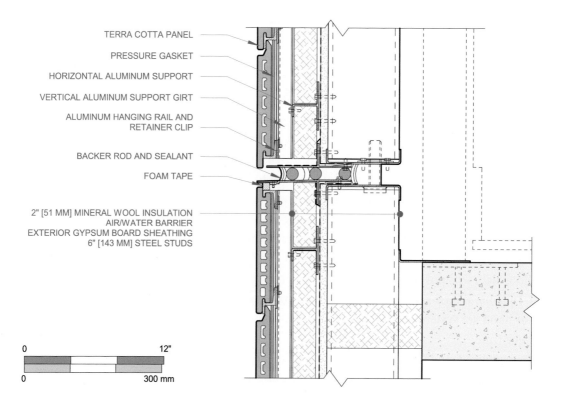

TERRA COTTA PANEL
PRESSURE GASKET
HORIZONTAL ALUMINUM SUPPORT
VERTICAL ALUMINUM SUPPORT GIRT
ALUMINUM HANGING RAIL AND RETAINER CLIP
BACKER ROD AND SEALANT
FOAM TAPE
2" [51 MM] MINERAL WOOL INSULATION
AIR/WATER BARRIER
EXTERIOR GYPSUM BOARD SHEATHING
6" [143 MM] STEEL STUDS

0   12"
0   300 mm

Figure 6.49 45 Province Street, Boston, Massachusetts (2009). Bruner/Cott Architects. Vertical section of stack joint in panelized metal stud walls, with rainscreen terra cotta applied off site. From drawings by Bruner/Cott and International Exterior Fabricators, Calverton, New York.

Figure 6.50 45 Province Street, Boston, Massachusetts. Left: Detail of cantilevered balconies emerging from behind solid wall planes. Right: Terminus of wall plane wrapped in terra cotta by Shildan Group, Mt. Laurel, New Jersey. Photos: Nat Crosby.

the sheet materials normally associated with the opaque zones of a curtain wall. Unitized systems offer rapid installation and high performance through gasketed, pressure equalized joints. The premium paid for terra cotta as the exterior finish is appropriate to an overall system that is of higher quality and price. The floor of the fabrication shop is an ideal place for setting rainscreen planks, and the common unit dimensions can be spanned with a continuous piece.

The work of Richard Meier and Partners, Architects (RMPA) often includes a featured wall plane that helps to organize the plan and contrasts with lighter or more transparent enclosure systems. At the Ara Pacis Museum in Rome and the Getty Center in Los Angeles, these walls are clad in travertine stone, quarried in Italy. The west façade of the United States Courthouse, San Diego (2012) achieves this gravitas with terra cotta on a unitized system. RMPA worked with NBK Keramik to develop the white clay body color the project required (Figure 6.52). The façade units were developed by Enclos, curtain wall specialists. The basic units are 10 feet

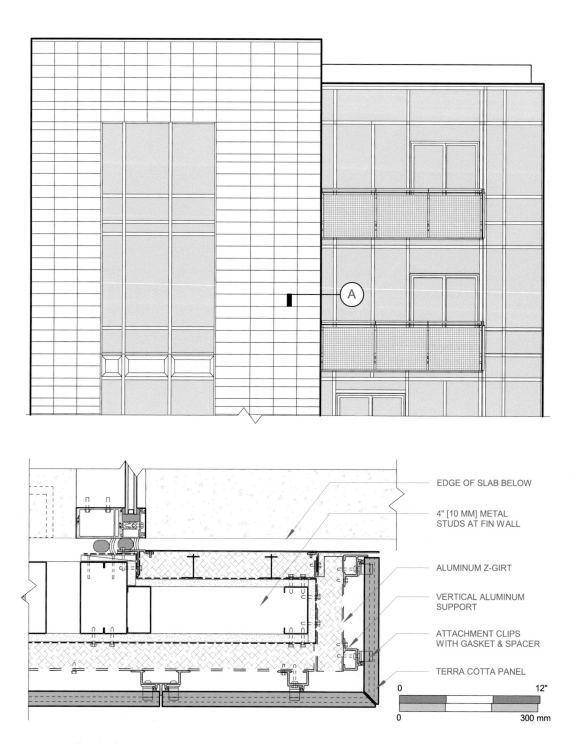

A

EDGE OF SLAB BELOW

4" [10 MM] METAL
STUDS AT FIN WALL

ALUMINUM Z-GIRT

VERTICAL ALUMINUM
SUPPORT

ATTACHMENT CLIPS
WITH GASKET & SPACER

TERRA COTTA PANEL

| 0 | 12" |
| 0 | 300 mm |

Figure 6.51 Plan detail and partial elevation of wall plane termination at residential balcony. 45 Province Street, Boston, Massachusetts. From drawings by Bruner/Cott and International Exterior Fabricators, Calverton, New York.

Figure 6.52 United States Courthouse, San Diego, California (2012). Richard Meier and Partners, Architects (RMPA). Photo: © Tim Griffith.

wide (3.05 m) and up to 23 feet tall (7 m), given the story heights required for courtrooms. In the section shown, there are offices for court staff against the façade with light taken over the top to the main spaces beyond (Figure 6.53). Terra cotta planks in 5 foot (1.52 m) lengths create strong horizontal bands, while integrated metal louver assemblies control solar gain at the glass.

The featured plane is expressed at the north where the solid, horizontal bands "fly by" the end of the building and return around the corner. This detail demonstrated the versatility of prefabrication as practiced by Enclos. The cantilevered elements are clad with terra cotta on all four faces, attached with a structural steel tube to the body of the building (Figure 6.54). Enclos describes these units as among the largest and heaviest they have ever built.

BSU Hamburg houses the Ministry for Urban Development and the Environment for the city and the state. Designed by Sauerbruch Hutton, it was the flagship of the International Building Exhibition in 2013. It is a high performance building with triple glazed windows, night ventilation, geothermal energy sources and no central air conditioning. The building is organized into seven linked "houses" at five stories and a central tower at twelve. The "houses" differentiate administrative departments and manage the scale of a 215,000 square foot (19,974 sqm) building. They are created by juxtaposing a deeply undulating façade on one side with a gently undulating face on the other, resulting in roughly triangular clusters. The complex shapes are enclosed with 2,044 unitized aluminum curtain wall panels by Gartner (Figure 6.55).

There are some straight wall segments, but a larger number that are either convex or concave, with two different radii. The standard units are 2.60 meters (8.53 ft) wide and 3.33 meters (10.92 ft) tall. As the daylight and energy flows were optimized, the percentage of glazing on the façades is much lower than one normally associates with unitized curtain walls. The flowing quality of the façade is strengthened by de-emphasizing the unit modules. In the window zone, powder coated aluminum sheets span from opening to opening, covering the unit joints. There are two terra cotta modules per unit, rather than a one to one match. The courses of glazed terra cotta hold the same color over multiple lengths, emphasizing the whole over the parts.

Bands of terra cotta are applied at the spandrel panel, at the transom bar, comprising a light shelf and external shading, and at a guardrail that

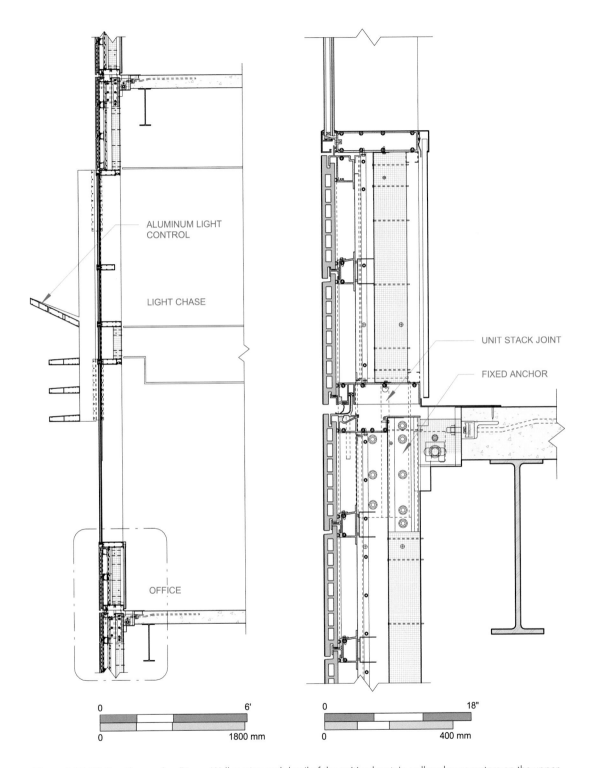

ALUMINUM LIGHT
CONTROL

LIGHT CHASE

OFFICE

UNIT STACK JOINT

FIXED ANCHOR

```
0                    6'
0            1800 mm
```

```
0                   18"
0             400 mm
```

Figure 6.53 US Courthouse, San Diego. Wall section and detail of the unitized curtain wall enclosure system on the upper levels of the west façade. Terra cotta by NBK. From drawings by RMPA and Enclos, façade contractors.

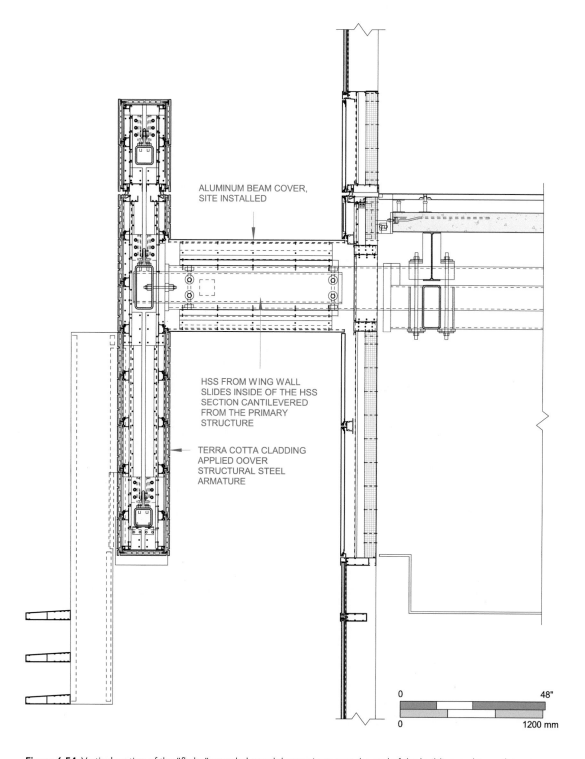

ALUMINUM BEAM COVER,
SITE INSTALLED

HSS FROM WING WALL
SLIDES INSIDE OF THE HSS
SECTION CANTILEVERED
FROM THE PRIMARY
STRUCTURE

TERRA COTTA CLADDING
APPLIED OOVER
STRUCTURAL STEEL
ARMATURE

0   48"

0   1200 mm

Figure 6.54 Vertical section of the "fly-by" spandrel panel that projects past the end of the building and turns the corner in front of the north façade. Prefabricated armature wrapped off site with terra cotta and attached using concentric, hollow steel tubes. US Courthouse, San Diego. From drawings by Enclos, façade contractors.

Figure 6.55 BSU Hamburg: Ministry for Urban Development and the Environment, Hamburg, Germany (2013). Sauerbruch Hutton, Architects. The undulating façade creates "houses" for administrative departments. Photo: Jan Bitter, DE.

prevents falls from the tall, tilt-turn windows. The rail is internally reinforced and back fastened where two terra cotta units span the window. There are vent flaps alongside the window that admit air through a wall cavity connected to the outdoors with perforated metal reveal panels at the jambs. This allows the building to be ventilated all night without concern for rain, wind or security (Figure 6.56).

Construction photographs indicate that some of the terra cotta panels were attached at the fabrication shop, some on the ground before lifting, and many were field applied after the curtain wall units were locked together. This last was particularly true of the curves. Standard shiplap joints were used where the connectors are not visible, but custom connections are moved to the back side near exposed edges. The extruded planks were post processed into true curves that must be matched by the outer edges of the metal heads and sills. The internal components of the wall are polygonal (Figure 6.57).

The terra cotta is finished with a semi-transparent glaze in twenty custom colors; five different shades of blue, red, yellow and green. The range of reds predominate the sunny façades, with blues, greens and yellows in the shade. The color mixes vary between the convex and concave portions and subtly transform as they move down the façades toward the tower.[20] All of the color distributions were laid out on the architectural drawings and radio frequency identity chips were used to make sure the tiles were installed in the proper places.[21]

Terra cotta facing on a precast concrete cladding panel can be the most cost-effective method for applying the material. The price range is comparable to hand set terra cotta, but it includes the structural backup wall as part of the package. There is an expectation that the costs will go down as subcontractors gain experience with the technique. Interior grade studs and gypsum finish may be added inside the shell, but they cost only 25% of the amount needed for an engineered, steel stud backup wall.[22]

Precast panels offer certain design freedoms. As the terra cotta is laid face down in the mold before concrete is poured, the pieces can vary a great deal in size, shape and pattern. Smaller pieces offer a technical advantage because there need be less concern about differential movement between the ceramic and the concrete. On the other hand, concrete panels themselves tend to be large, simple in shape, and not as nimble as unitized curtain walls.

The bond between concrete and ceramic products can be challenging. Developing a chemical bond between concrete and brick is assisted by the porosity and absorptivity of the brick. Terra cotta has lower values in both areas. The chemical bond can be designed to hold the tiles securely with

[20] NBK Keramik GmbH, "BSU Hamburg."
[21] Gartner.
[22] Streff.

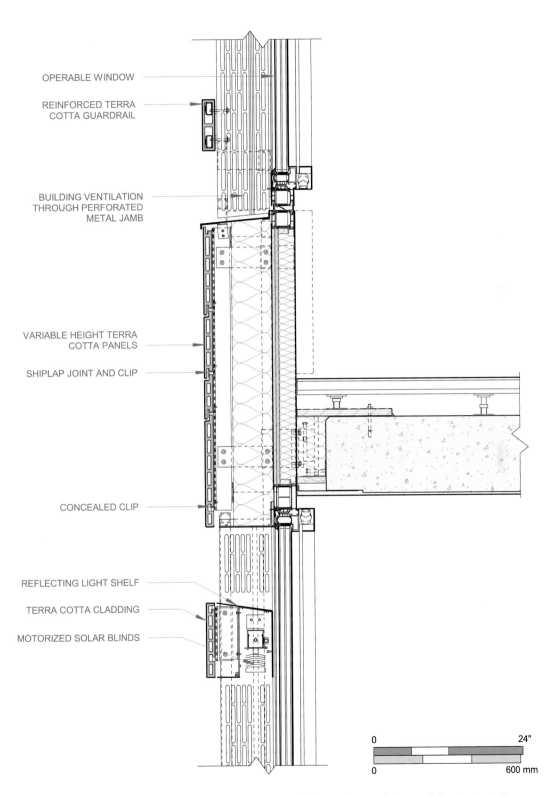

OPERABLE WINDOW

REINFORCED TERRA
COTTA GUARDRAIL

BUILDING VENTILATION
THROUGH PERFORATED
METAL JAMB

VARIABLE HEIGHT TERRA
COTTA PANELS

SHIPLAP JOINT AND CLIP

CONCEALED CLIP

REFLECTING LIGHT SHELF

TERRA COTTA CLADDING

MOTORIZED SOLAR BLINDS

0   24"

0   600 mm

Figure 6.56 BSU Hamburg. Unitized curtain wall with terra cotta cladding at the guardrail, spandrel and light shelf. Perforated metal jamb liners admit ventilation air though operable panels beside the window. From drawings provided by Sauerbruch Hutton, Architects.

Figure 6.57 BSU Hamburg. Above: Gently curved, extruded tiles prepared for installation. Below: Curtain wall unit on site with metal head and sill to meet tiles. Photos: Frank Kaltenbach, Munich, Germany.

attention to the porosity of the clay body. Mechanical keys can be formed into the back of the clay to provide a factor of safety.

Profiles intended for bonded facing are extruded in closed forms with weak points in them so that the back side can be broken away without damaging the front (see Figure 3.19). Folded surfaces increase the contact area available for bonding when filled with concrete. Deceptively simple exterior forms may require an intricate web to stabilize the shape through drying and firing. As the webs are cut away, they leave keys to lock with the concrete backing (Figure 6.58).

Henry W. Bloch Executive Hall (2013) is the signature building for the School of Management at the University of Missouri – Kansas City. It was designed by BNIM and Moore Ruble Yudell, Architects, to reintegrate an eclectic campus. They chose a simple exterior form to be clad in glazed terra cotta with colors that derive from the brick and stone masonry of adjacent buildings (Figure 6.59). The long life and low maintenance of terra cotta were attractive, but the cost of a hand set rainscreen system exceeded the project budget. BNIM had experience with internally insulated, precast concrete wall systems on other projects. The continuous layer of insulation provides good energy performance with stabilizing thermal mass added by the internal leaf of concrete. Applying the terra cotta at the precast site greatly reduced the cost. The layout of colors was carefully considered and fully documented on the elevations without an algorithm or repetitive modules.

The façades were laid out on a 4 foot (1.23 m) module, corresponding to the length of the terra cotta planks. The concrete panels are 12 feet (3.66 m) wide and full story heights of 16 to 18 feet (4.88–5.49 m). Careful stacking of the window openings preserved the structural integrity of the panels. In the few places where infill spandrel panels were needed, they could be laterally stabilized through attachment to the adjacent panels (Figure 6.60).

In order to maintain the layout grid of the elevations and reconcile the exterior wall thickness, a particularly inventive detail was needed at the outside corner (Figure 6.61). The warm side leaf of concrete was mitered to close at the interior corner. Continuity of the insulation was achieved from the exterior, then covered with a self-adhesive barrier. The corner was stitched shut with custom formed terra cotta elements that match the face profile of the adjacent planks and were attached with conventional rainscreen clips.

SPECIAL CONDITIONS

Terra cotta can be cut very cleanly and precisely. For a wall clad in open jointed rainscreen planks, the most logical and economical method to

Figure 6.58 Palagio Engineering, Greve in Chianti (Florence), Italy. Fabrication of terra cotta facing for precast concrete components. Left: Completed segment, trimmed and glazed. Right: Sections as extruded, with internal webs to stabilize the form during drying and initial firing. Photos: Donald Corner.

Figure 6.59 Henry W. Bloch Executive Hall, Kansas City, Missouri (2013). Designed by BNIM and Moore Ruble Yudell, architects. Above: East entry elevation. Below: custom closure piece on southwest corner. Terra cotta clad, precast concrete panels. Photos: Ashley Streff.

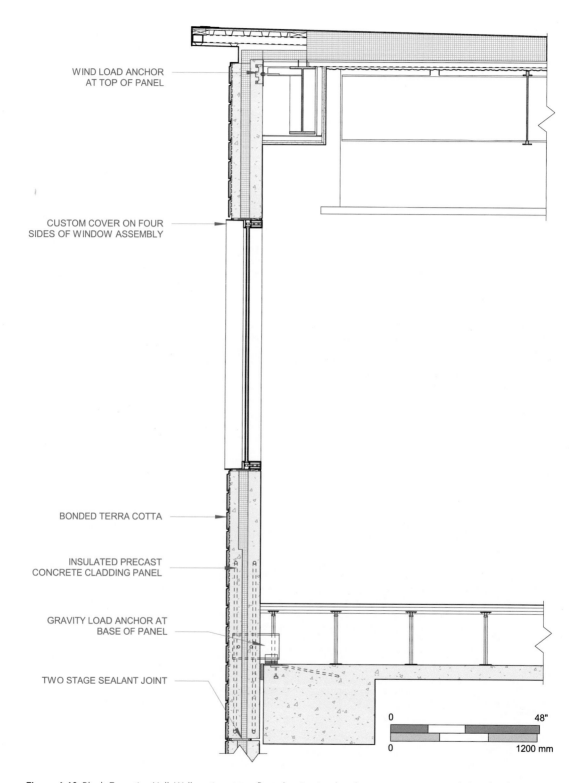

WIND LOAD ANCHOR
AT TOP OF PANEL

CUSTOM COVER ON FOUR
SIDES OF WINDOW ASSEMBLY

BONDED TERRA COTTA

INSULATED PRECAST
CONCRETE CLADDING PANEL

GRAVITY LOAD ANCHOR AT
BASE OF PANEL

TWO STAGE SEALANT JOINT

| 0 | | 48" |
| 0 | | 1200 mm |

Figure 6.60 Bloch Executive Hall. Wall section at top floor showing insulated, precast concrete panels faced with terra cotta extrusions by NBK. From drawings by BNIM, Architects, Kansas City, Missouri.

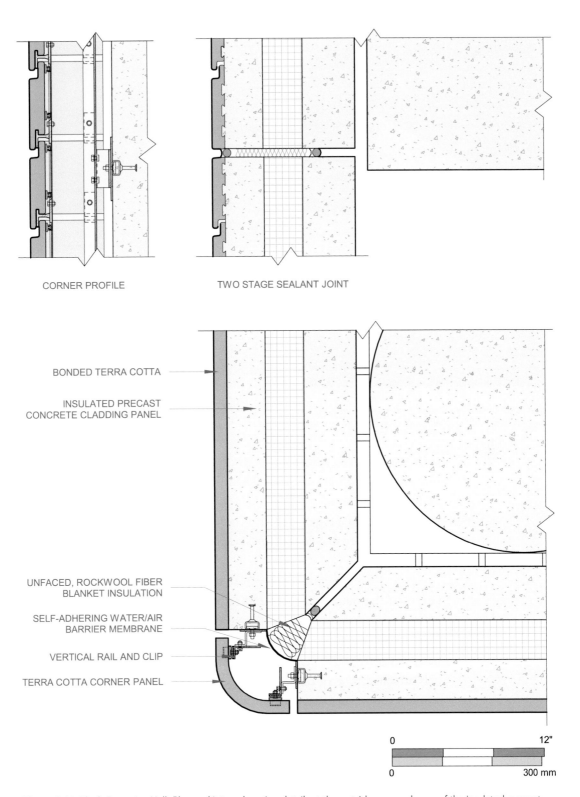

CORNER PROFILE

TWO STAGE SEALANT JOINT

BONDED TERRA COTTA

INSULATED PRECAST
CONCRETE CLADDING PANEL

UNFACED, ROCKWOOL FIBER
BLANKET INSULATION

SELF-ADHERING WATER/AIR
BARRIER MEMBRANE

VERTICAL RAIL AND CLIP

TERRA COTTA CORNER PANEL

0   12"

0   300 mm

Figure 6.61 Bloch Executive Hall. Plan and internal section details at the outside corner closure of the insulated, precast wall panels. Custom molded corners match profiles of the bonded, shiplap panels. From drawings by BNIM and NBK Architectural Terra Cotta.

Figure 6.62 Li Ka Shing Center for Biomedical and Health Sciences, University of California, Berkeley (2011). Zimmer Gunsel Frasca, Architects, Portland, Oregon. Miter cut terra cotta rainscreen corners by NBK. Photos: Donald Corner.

turn a corner is to simply miter the two adjacent planks (Figure 6.62). If one side of the corner is shorter than the other, it may not be desirable to hand set such a small piece. A single corner element can be created by gluing the miter cuts together. This is best done at the point of manufacture with the capacity to color match the epoxy adhesive with the glazed finish (Figure 6.63).

Renzo Piano's seminal project at Lodi is finished with short radius curves at the outside corners and returns. Executed at a time when the entire façade was made up from smaller parts, the corner consists of two. One has the radius and a false joint to make the assembly appear diagonally symmetrical. Where the corner aligns with the typical wall cladding, it is attached to a vertical fin in the same manner (see Figure 6.2). The window returns are suspended from pins that directly engage the concrete structure (Figure 6.64). Piano's iconic project at Potsdamer Platz followed with rounded corners, formed out of a single piece of extruded, horizontal profile (see case study of Chapter 9).

The Löwenbräu Areal in Zurich includes a new office building designed by Gigon-Guyer Architekten. The wavy red façade is composed of vertically oriented profiles, coursed to fit the window pattern (Figure 6.65). The strict

Figure 6.63 M9 Museum, Venice Mestre (Venice), Italy (2019). Epoxy bonded right angle corners with color match to the terra cotta glaze. Fabricated by NBK Keramic, Emmerich, Germany. Sauerbruch Hutton, Architects. Photos: Donald Corner.

modularity of the scalloped shapes is maintained at the corner with a piece designed to fit within the area limits of an extrusion die (Figure 6.66).

One-piece corners with an intermediate radius can be achieved in horizontal cladding through post processing the extruded section (see Figure 3.17). Slump forming or hand forming into molds may be used. Previously fired ceramic shapes may be needed to support the curved shape in the kiln. Wood forms are used to support the work while end cuts are made to the correct circumference and angle. Large radius curves are more readily achieved in either the horizontal orientation as at BSU Hamburg (see Figure 6.55), or the vertical as in the Roberts Pavilion (see Figure 6.46).

Hollow core, extruded terra cotta sections are intrinsically volumetric. As they are most often mounted parallel to the façade plane, they demonstrate three-dimensionality through a folded, curved or rippled surface profile. A bolder expression can be attained by turning the long axis of a section so that it is perpendicular to the surface of the building.

Figure 6.64 Banca Popolare di Lodi, Italy. Two part, terra cotta corners with false joints to articulate the quarter radius portion. Fabricated by Palagio Engineering. Renzo Piano Building Workshop, architects. Photos: Donald Corner.

Figure 6.65 Löwenbräu Areal, Zurich, Switzerland (2014). East office building. Gigon-Guyer Architekten. Vertically oriented, fluted terra cotta profiles by NBK. Photo: René Dürr Fotograf, Zurich.

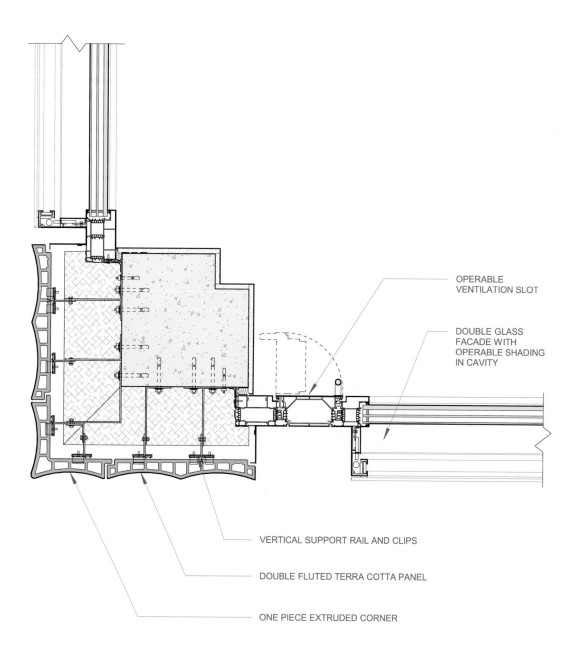

OPERABLE
VENTILATION SLOT

DOUBLE GLASS
FACADE WITH
OPERABLE SHADING
IN CAVITY

VERTICAL SUPPORT RAIL AND CLIPS

DOUBLE FLUTED TERRA COTTA PANEL

ONE PIECE EXTRUDED CORNER

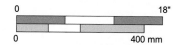

Figure 6.66 Plan detail at outside corner. Löwenbräu Areal, Zurich. Custom corner piece with dimensions that can be accommodated within a single extrusion die. Fabricated by NBK Keramik. From drawings by Gigon-Guyer Architekten.

This can be done with baguettes at a small scale by applying clips, differentially, to the cores of the extrusion. The community library at Greve in Chianti (see Figure 6.28) and the Kirkland Museum (see Figure 6.43) provided examples of this approach. At larger scales, the rotated section can produce a substantial vertical fin with distinct visual impact as well as shading potential. The systems of attachment must also be scaled up to resist lateral loads due to wind or impact, which act to twist the cantilevered fin.

One Crown Place, in the London borough of Hackney, is a regenerated block with two towers added to a mix of uses near the ground. KPF Architects retained several existing buildings and organized the new construction around a central courtyard. The towers were developed through parametric modeling to optimize daylight access and views, while protecting the scale of the heritage block. With rotated, polygonal plan shapes, the tower façades are viewed obliquely. Projecting fins add interest to this perspective.

Terra cotta cladding was selected to complement the brick, glazed brick and ceramic tile found in the context. The color range of red, orange and brown responds to the utilitarian brick colors of the borough. The projecting fins and adjacent ribbed panels rise continuously over two-story intervals, interrupted by a projecting spandrel cover of powder coated aluminum (Figure 6.67).

Figure 6.67 One Crown Place, London, UK. Kohn Pederson Fox Architects (2020). Mixed use development. Residential towers clad in terra cotta with perpendicular fins. Palagio Engineering. Photo: © Agnese Sanvito.

Both profiles are attached to unitized curtain wall modules. Aluminum extrusions were silicone bonded to the first cell of the fins at the point of manufacture, Palagio Engineering. This begins the interface between the fin and the wall. The silicone is able to cushion differential movements between the materials. The connection is completed with additional splines and struts that ultimately attach to the side of a unit joint. The rectangular ribs complement the fin and are configured to block views through the vertical joints, which is particularly important near the ground (Figure 6.68).

One Great Jones Alley (2019) is a mixed use building in New York, named for the residential entrance at the rear. The primary façade faces west at 688 Broadway, in the NOHO Historic District. Buildings on this part of the street have solid masonry corners with multiple windows clustered in the centers. At 700 Broadway is an 1891 building by George B. Post with two tiers of multistory arcades in the façades, supported by deep masonry piers. Responding to this context, BKSK Architects gave the new building large scale vertical louvers made with extruded baguettes. Appropriate for late afternoon shading, the louvers also give layers to a façade that is seen obliquely along a well defined urban street.

The asymmetrical terra cotta elements are 18 inches (45.72 cm) deep, in segments up to 4 feet (1.23 m) in height. They are arranged in two-story tiers with a strong expression of alternate floor lines and partially concealed floors in between (Figure 6.69). The louver segments are stacked and stabilized against rotation by two extruded aluminum armatures concealed in the cells. These are fixed to end plates and tied back to the concrete structure, passing through insulated sheet metal spandrel covers (Figure 6.70).

The House of Music, Innsbruck, Austria (2018), features louvers that are assembled as a composite of smaller extruded sections (Figure 6.71). The building was designed by Erich Strolz and Dietrich Untertrifaller Architekten. It is a prismatic box, centered in a parkscape, surrounded by white and yellow historic structures. The façades are composed in three distinct zones. The large music performance volumes in the middle of the building are clad in varying vertical profiles, finished with an iridescent glaze that shifts from reddish brown to almost black, depending on sky conditions. Public areas near the ground are fully glazed. Higher in the building there is a complex arrangement of small and intermediate spaces, with horizontal runs of windows in the vision zone. These areas are unified and partially concealed by tiers of full story vertical louvers that are faced in terra cotta with the same glaze.[23]

23 *Detail*, pp. 54–61.

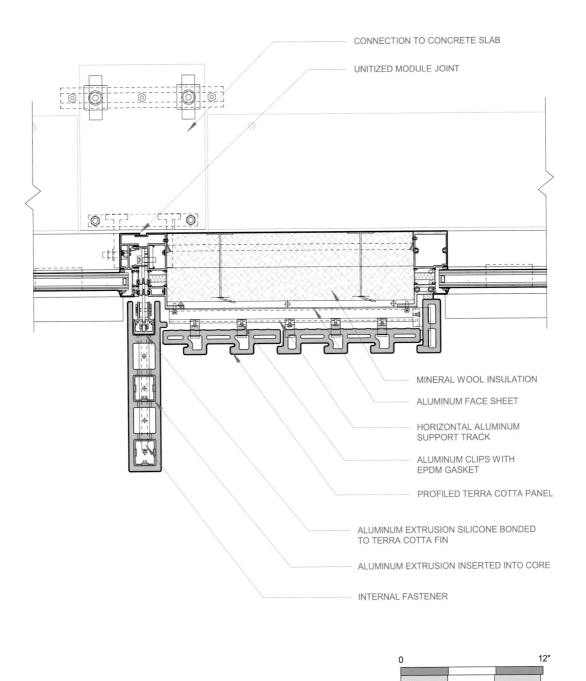

CONNECTION TO CONCRETE SLAB

UNITIZED MODULE JOINT

MINERAL WOOL INSULATION

ALUMINUM FACE SHEET

HORIZONTAL ALUMINUM
SUPPORT TRACK

ALUMINUM CLIPS WITH
EPDM GASKET

PROFILED TERRA COTTA PANEL

ALUMINUM EXTRUSION SILICONE BONDED
TO TERRA COTTA FIN

ALUMINUM EXTRUSION INSERTED INTO CORE

INTERNAL FASTENER

0		12"
0		300 mm

Figure 6.68 Plan detail of the unitized curtain wall at One Crown Place. Perpendicular terra cotta fins attached to the edge of one façade unit, at the meeting point with another. From Drawings provided by KPF Architects and Scheldebouw, Middelburg, Netherlands.

Four plane sections are gathered around an aluminum core to create a foil shape in plan, using only simple, repetitive elements. The resulting shape is stiffened in the cross-axial direction with a solid bar at the center. The louvers are attached top and bottom using cylindrical axels mounted

Figure 6.69 One Great Jones Alley, 688 Broadway, New York, NY. BKSK Architects, New York (2019). Large-scale vertical louvers by Boston Valley Terra Cotta. Photo: Field Condition.

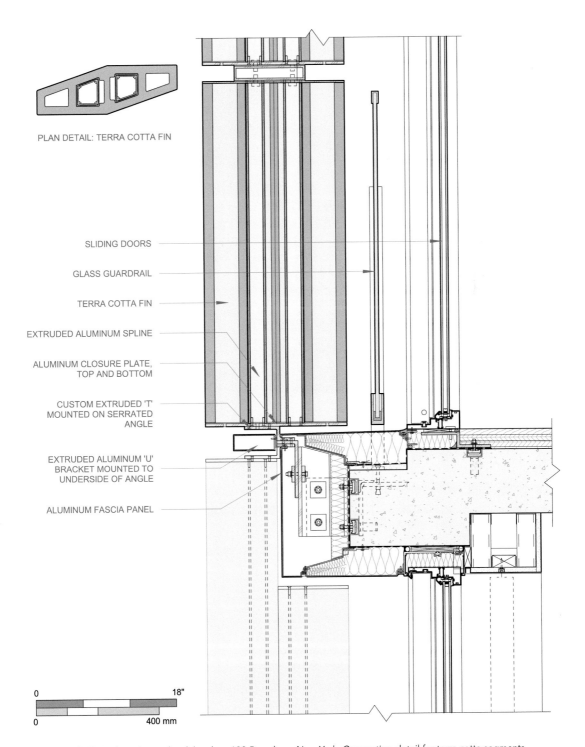

PLAN DETAIL: TERRA COTTA FIN

SLIDING DOORS

GLASS GUARDRAIL

TERRA COTTA FIN

EXTRUDED ALUMINUM SPLINE

ALUMINUM CLOSURE PLATE,
TOP AND BOTTOM

CUSTOM EXTRUDED 'T'
MOUNTED ON SERRATED
ANGLE

EXTRUDED ALUMINUM 'U'
BRACKET MOUNTED TO
UNDERSIDE OF ANGLE

ALUMINUM FASCIA PANEL

0   18"

0   400 mm

Figure 6.70 Vertical section at the slab edge, 688 Broadway, New York. Connection detail for terra cotta segments joined by extruded aluminum splines. BKSK Architects. Boston Valley Terra Cotta. From drawings by Curtain Wall Design Consultants, Dallas, Texas.

Figure 6.71 House of Music, Innsbruck, Austria (2018). Erich Strolz and Dietrich Untertrifaller Architekten. Clad with vertical terra cotta profiles and composite louvers by NBK Keramik. Photo: Roland Halbe.

on horizontal box sections. This mechanism allows ranks of louvers to be positioned at different angles depending on the external exposure and the internal space requirements. The building gains a strong vertical expression to complement the horizontal zones of program spaces and roof terraces. To differing degrees, activities in the complex are concealed or exposed through the deeply colored façade (Figure 6.72).

Aluminum armatures allow readily fabricated terra cotta sizes and shapes to be added together to make larger scale architectural elements. These compound elements may be attached to the wall plane, or free standing. Examples of both are pictured in the case study of 125 Deansgate in Chapter 9.

Given autonomy by their internal structure, larger scale composite elements can also be applied in more than one layer. This brings even greater depth of light and shadow to a façade. BKSK Architects exploited this opportunity for the Gatehouse to One Madison (2017). The building provides an elegant, remote entry to a residential tower located at the south end of Madison Avenue. Fronting on a relatively quiet block of East 22nd Street, the entry building complements the neighboring buildings that have window planes set deeply within volumetric stone façades. At the Gatehouse, glass is withdrawn behind deep column and spandrel covers made with combinations of flat and custom profiled terra cotta (Figure 6.73). The column covers have continuous, external corners extruded as a single "L" shape with faces of 15 and 9.5 inches (38 and 24 cm). A central rib projects from the column face and this form is picked up in multiple layers of vertical fins that screen the south facing glass. The outward faces of the fin are formed in a single extrusion, with custom flat pieces used to close the backs and ends where they are exposed (Figure 6.74). The glaze is a warm cream color with variations that add richness beyond that provided by reflected sunlight and shadow. A layered strategy composed of smaller elements, woven together, can be seen in the case study of Central Saint Giles in Chapter 9.

One particular area of detail that demonstrates the range of terra cotta applications occurs at the junction of the window and the wall. In highly rationalized, contemporary rainscreen applications, this junction is often fulfilled by the window supplier. The case study of Tykeson Hall, in Chapter 9, illustrates a direct and efficient approach. There, the snap-on cover of a pressure plate glazing system is replaced with a custom aluminum profile that closes off the cavity behind the rainscreen surface.

Looking back to historical examples, shaped terra cotta was often used to create a zone of transition between the window and primary wall

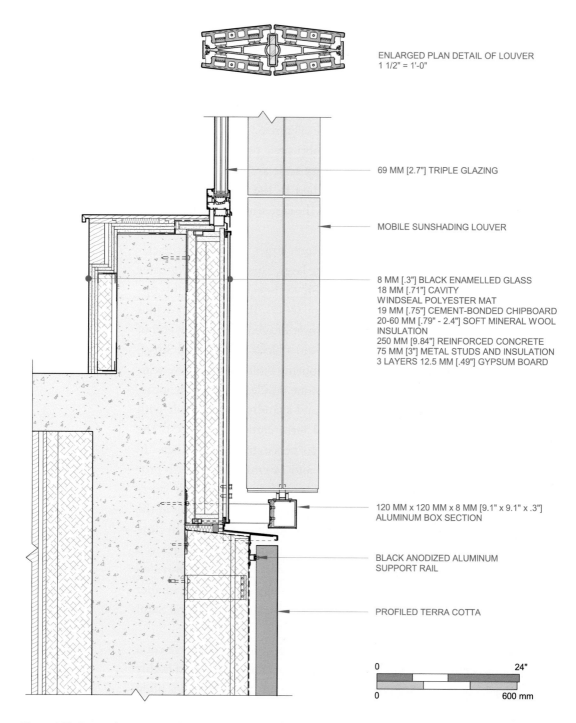

ENLARGED PLAN DETAIL OF LOUVER
1 1/2" = 1'-0"

69 MM [2.7"] TRIPLE GLAZING

MOBILE SUNSHADING LOUVER

8 MM [.3"] BLACK ENAMELLED GLASS
18 MM [.71"] CAVITY
WINDSEAL POLYESTER MAT
19 MM [.75"] CEMENT-BONDED CHIPBOARD
20-60 MM [.79" - 2.4"] SOFT MINERAL WOOL
INSULATION
250 MM [9.84"] REINFORCED CONCRETE
75 MM [3"] METAL STUDS AND INSULATION
3 LAYERS 12.5 MM [.49"] GYPSUM BOARD

120 MM x 120 MM x 8 MM [9.1" x 9.1" x .3"]
ALUMINUM BOX SECTION

BLACK ANODIZED ALUMINUM
SUPPORT RAIL

PROFILED TERRA COTTA

| 0 | 24" |
| 0 | 600 mm |

Figure 6.72 House of Music. Vertical section and partial plan detail. Shading louvers composed of four extruded sections attached to a central armature. Axle permits adjustment of the shading angles. From drawings by Dietrich Untertrifaller Architekten and Starmann Metallbau, Klagfurt, Austria.

materials, whether brick or stone. Capitalizing on the economics of repetition, terra cotta could be given ornamental detail at this critical junction, without extraordinary cost. The city hall of Oakland, California

Figure 6.73 Gatehouse to One Madison, New York, NY (2017). BKSK Architects. Terra cotta column and spandrel covers with vertical fins in multiple layers. Boston Valley Terra Cotta. Photo: Raimond Koch.

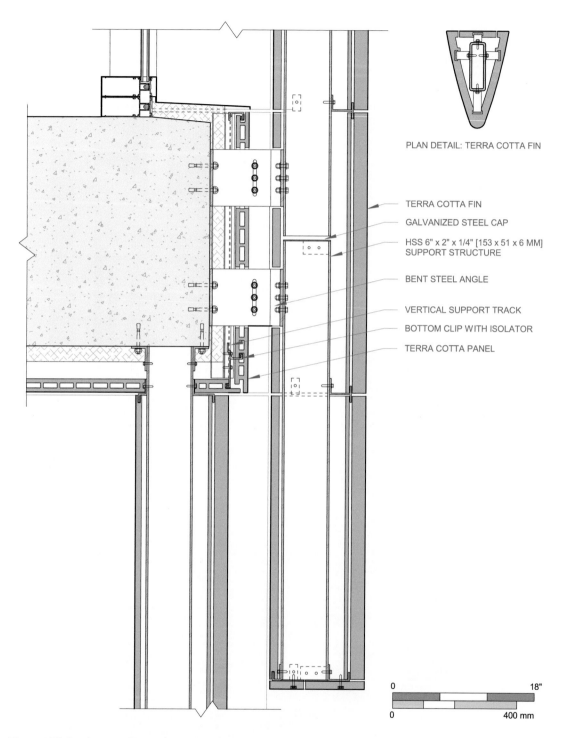

PLAN DETAIL: TERRA COTTA FIN

TERRA COTTA FIN

GALVANIZED STEEL CAP

HSS 6" x 2" x 1/4" [153 x 51 x 6 MM]
SUPPORT STRUCTURE

BENT STEEL ANGLE

VERTICAL SUPPORT TRACK

BOTTOM CLIP WITH ISOLATOR

TERRA COTTA PANEL

| 0 | | 18" |
| 0 | | 400 mm |

Figure 6.74 Gatehouse to One Madison. Vertical section at second floor slab edge. Terra cotta fins assembled around hollow steel tubes. BKSK Architects. From shop drawings by Boston Valley Terra Cotta.

Figure 6.75 Left: City Hall, Oakland, California (1914). Designed by Palmer & Hornbostel of New York. Terra cotta ornament at the windows combined with smooth blocks of granite. Photo: Donald Corner. Right: 14th Street Development, Washington, DC (2019). Selldorf Architects. Glazed terra cotta window reveals combined with limestone. Photos: CORE Architecture + Design.

(1914) is an example. The steel structure is clad with blocks of Sierra granite, a durable material that does not lend easily to carving.[24] Cream colored, terra cotta elements provide detail around and within the window groupings using classical ornament and motifs from California agriculture (Figure 6.75, left image).

Selldorf Architects reinterpreted this tradition at a mixed-use development on 14th Street in Washington, DC (2019). Fitting new, seven-story buildings between two existing structures on the site resulted in structural bays of different dimensions. The architects decided to hold the window modules and the limestone cladding as constants to unify the complex (Figure 6.75, right image). The differences were taken up by changing the angles of the window reveal. In this case, terra cotta was chosen for brilliant colors rather than ornament (Figure 6.76).

[24] Kurutz, p. 105.

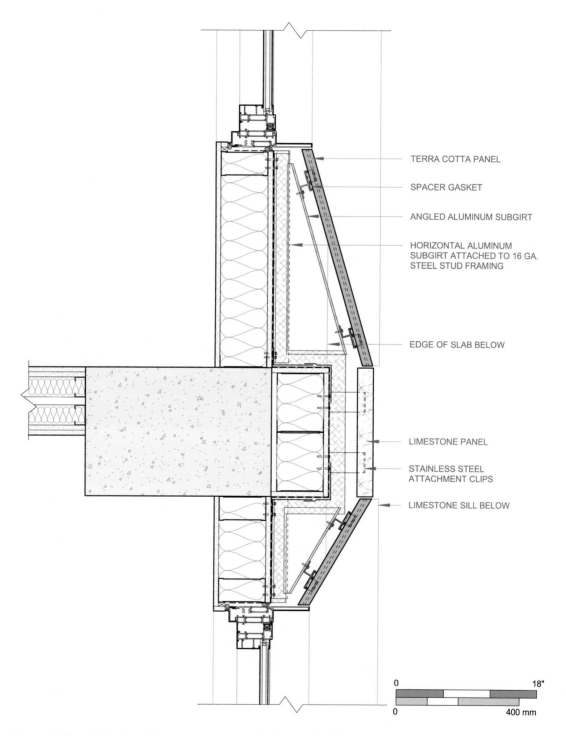

TERRA COTTA PANEL

SPACER GASKET

ANGLED ALUMINUM SUBGIRT

HORIZONTAL ALUMINUM
SUBGIRT ATTACHED TO 16 GA.
STEEL STUD FRAMING

EDGE OF SLAB BELOW

LIMESTONE PANEL

STAINLESS STEEL
ATTACHMENT CLIPS

LIMESTONE SILL BELOW

Figure 6.76 Plan detail of the 14th Street Development, Washington, DC (2019). Limestone panels cover the concrete columns while glazed terra cotta window reveals accommodate variation in the bay spacing. From drawings by Selldorf Architects and CORE Architecture + Design, Washington, DC.

The South Molton Street Building (2012) is located just off Oxford Street in Mayfair, London (Figure 6.77). DSDHA, Architects, designed the building to be completed on a nine-month schedule in anticipation of the Olympics. Adjacent buildings are clad in brick with carved stone window mullions and wood sash. DSDHA responded to the context using terra cotta to fill the roles of all the preceding materials. The building reflects its construction with vertical terra cotta profiles lying parallel to the steel stud walls that span from floor to floor. The varied profiles are intercepted with box aluminum horizontals to pick up critical datum lines from the abutting façade. This allows the terra cotta elements to shift in position and sequence from one tier to the next, controlling the scale of the building (Figure 6.78).

The asymmetrical, rippling profiles were inspired by the historic course of the River Tyburn flowing along Molton Street.[25] Broader shapes cover the solid portions of the façade. They are attached in a conventional fashion, except at the boundary where the last tile cantilevers over the window mullion. The intersections between tiles overlap at an angle so that there is no direct line of sight to the rails behind. Where the mullions are freestanding, slender baguettes are attached directly to the face of the pressure plate using clips anchored in a tubular spline, bonded inside the cells of the clay extrusion. The façade, as a whole, curves around the end to make the sharp angle at Davies Street. As the façade turns, the pressure plate that supports the baguettes is silicone bonded to the glass (Figure 6.79).

The office building at 125 Deansgate, Manchester, develops larger, vertical profiles with a composition of shapes attached to the front of a full curtain wall unit, rather than just a single mullion. The adjacent vision unit uses structural silicone glazing to conceal the aluminum and bring the glass to the edge of the terra cotta. This project is described further in the case studies of Chapter 9 (see Figure 9.4.2).

The intersection between wall and window plays across the entire façade at 529 Broadway, New York (1916) by BKSK Architects. The site was once occupied by the Prescott Hotel of 1852. Down Spring Street, at the far end of the site, a complementary annex to the hotel still stands. It has a masonry façade, punched by a grid of rectangular openings with ornamental lintels. One door beyond, the Donald Judd Foundation occupies an 1870s building that he purchased. It has a highly glazed cast iron front. The concept for the new building was to memorialize the transformation of

[25] DSDHA, Architects.

Figure 6.77 South Molton Street Building, London, UK (2012). DSDHA Architects. Vertically oriented terra cotta profiles cover the window wall mullions and directly abut the glass. NBK Keramik. Photo: John Rowell.

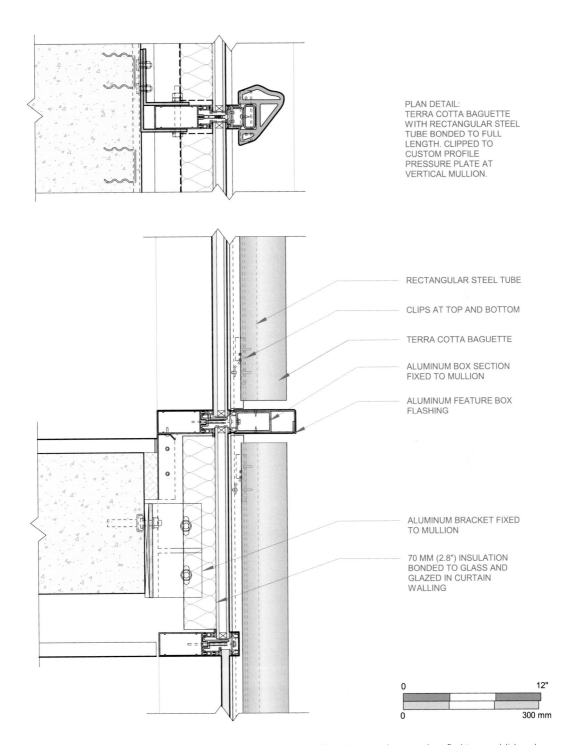

PLAN DETAIL:
TERRA COTTA BAGUETTE
WITH RECTANGULAR STEEL
TUBE BONDED TO FULL
LENGTH. CLIPPED TO
CUSTOM PROFILE
PRESSURE PLATE AT
VERTICAL MULLION.

RECTANGULAR STEEL TUBE

CLIPS AT TOP AND BOTTOM

TERRA COTTA BAGUETTE

ALUMINUM BOX SECTION
FIXED TO MULLION

ALUMINUM FEATURE BOX
FLASHING

ALUMINUM BRACKET FIXED
TO MULLION

70 MM (2.8") INSULATION
BONDED TO GLASS AND
GLAZED IN CURTAIN
WALLING

0   12"

0   300 mm

Figure 6.78 Vertical section at a floor slab, South Molton Street Building. Exterior aluminum box flashing establishes datum lines. Extruded terra cotta profile covers window mullion. From drawings by DSDHA and Procare Facades, Desborough, Northamptonshire, UK.

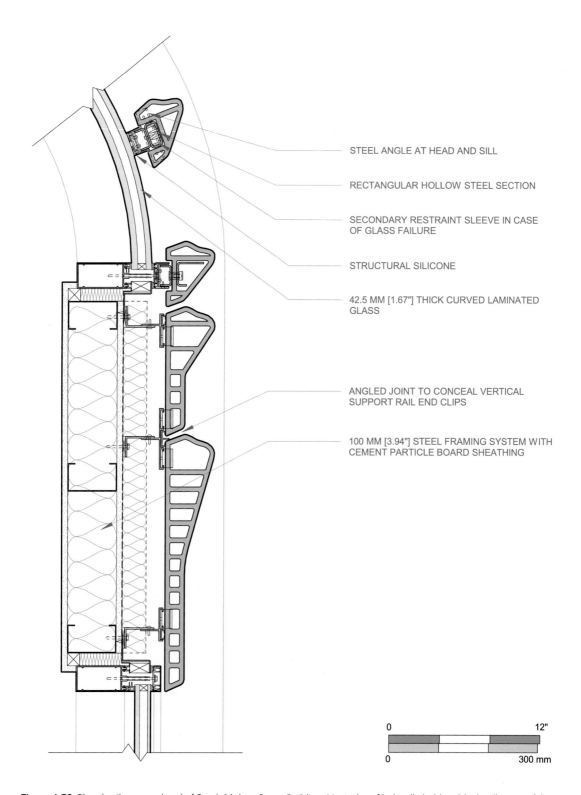

STEEL ANGLE AT HEAD AND SILL

RECTANGULAR HOLLOW STEEL SECTION

SECONDARY RESTRAINT SLEEVE IN CASE OF GLASS FAILURE

STRUCTURAL SILICONE

42.5 MM [1.67"] THICK CURVED LAMINATED GLASS

ANGLED JOINT TO CONCEAL VERTICAL SUPPORT RAIL END CLIPS

100 MM [3.94"] STEEL FRAMING SYSTEM WITH CEMENT PARTICLE BOARD SHEATHING

0   12"

0   300 mm

Figure 6.79 Plan detail at curved end of South Molton Street Building. Vertical profiled wall cladding blocks all views of the backup structure and window mullions. Additional profiles attached directly to the curved glass. From drawings by DSDHA and Procare Facades, Desborough, Northamptonshire, UK.

enclosure technologies. The Spring Street elevation begins with rectangular openings, scaled to its immediate neighbor. The grain of the brickwork and the ornament at the windows are both realized in custom cast terra cotta, applied as a rainscreen in front of a unitized curtain wall (Figure 6.80).

Moving toward Broadway, the spandrel panels begin to twist out of plane and fold back on themselves, revealing more of the curtain wall. The column covers also reduce in size progressively, reaching the slender proportions of the cast iron precedent. Fritted glass, spandrel glass and secondary mullions are used to retain the original size of the view openings. As the façade turns the corner, it is a contemporary glass wall system overlaid with projecting fins that preserve the scale and character of the blocks in SOHO.

The techniques used in this transforming façade are a thorough update of those developed at Lodi, by Renzo Piano. The terra cotta units were slip cast in relatively small units to generate the overall rotation without twisting each piece. Digital tools were used to capture the ornamental motifs from historic photographs, as well as to generate the progression of forms. Film-making software was used to digitally "cast" 700 distinct positives that led to the creation of molds.

The many pieces are back fastened to horizontal carriers using a variety of hook plates and bolted armatures. The carrying tubes fit over shear blocks that are mounted to the sides of custom fins emerging from within the curtain wall mullions. In the first stage of the progression, the fin cantilevers upward, past the stack joint at the floor line. This allows the closely spaced pieces of the spandrel panel to be tied to just one façade unit (Figure 6.81).

ADDITIONAL FORMING TECHNIQUES

Ram pressed terra cotta units can vary in size from what might be considered a wall tile up to large-scale architectural panels. The choice of pressing as a forming method implies a three-dimensional surface that is not possible with extrusion. The elevation of a single unit need not be rectangular. Molds can be designed to produce any polygonal shape, usually those capable of interlocking for efficient coverage of a surface (Figure 6.82). Simple pressed forms are hollowed out at the back to reduce weight and provide a relatively consistent clay thickness to facilitate drying and firing. A pressed unit can vary greatly in depth, measured perpendicular to the wall plane. This section presents a sequence of examples ranging from surface variations to full, prismatic volumes.

Figure 6.80 529 Broadway, New York, NY (2016). BKSK Architects. Above: The corner of Spring Street (left) and Broadway. Below: Spandrel panels on Spring Street rotate out of plane and reduce to projecting, horizontal fins with extensive glazing as the curtain wall units approach Broadway. Terra cotta by Sannini, Impruneta (Florence), Italy. Photos: © Christopher Payne/Esto.

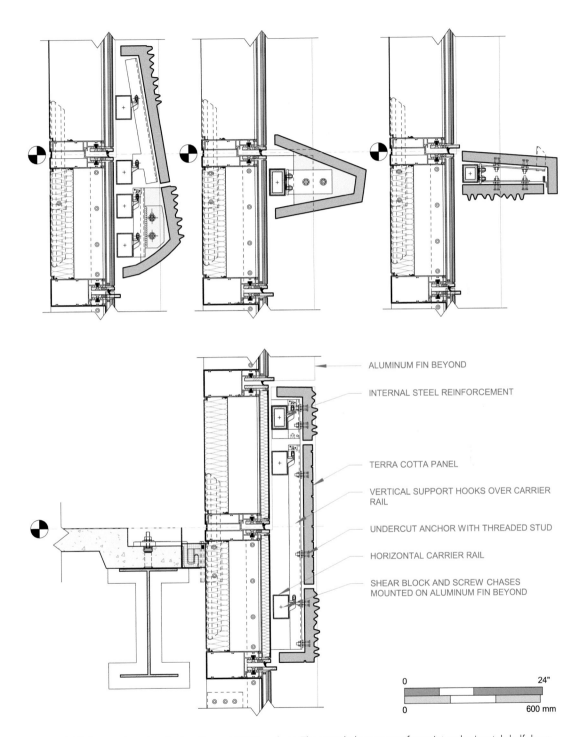

ALUMINUM FIN BEYOND

INTERNAL STEEL REINFORCEMENT

TERRA COTTA PANEL

VERTICAL SUPPORT HOOKS OVER CARRIER RAIL

UNDERCUT ANCHOR WITH THREADED STUD

HORIZONTAL CARRIER RAIL

SHEAR BLOCK AND SCREW CHASES MOUNTED ON ALUMINUM FIN BEYOND

0   24"

0   600 mm

Figure 6.81 Progression of vertical sections at 529 Broadway. The spandrel cover transforms into a horizontal shelf down the length of the Spring Street elevation. Short lengths of terra cotta are attached to a carrier system that is part of the curtain wall unit installed below the floor level. From drawings by BKSK and Front, Inc., façade specialists.

Figure 6.82 Ram pressed wall tiles produced for 11 Beech Street, New York, NY (2018). BKSK Architects. Palagio Engineering, Greve in Chianti (Florence), Italy. Photo: Donald Corner.

The Trumpf Campus Restaurant in Ditzingen, Germany is a pavilion structure with a distinctive, overhanging roof plate. Viewed from underneath, it is a leaf-like web of polygonal cells formed by intersecting planes of glue laminated timber. Barkow Leibinger Architects of Berlin worked with the creative engineer Werner Sobek to develop the framing system, which is the product of digital representation and fabrication (Figure 6.83).

To complement the geometric character of the roof, the solid walls are clad in ram pressed terra cotta shapes both inside the building and out. The units have an asymmetrical kite shape derived by slicing away two opposite corners from a 35 cm (1.15 ft) square. With a constant thickness at the boundaries, the kites are pressed into convex and concave versions with even more asymmetrical ridges and valleys.

The units fall between big tiles and small panels. On the interior, they are adhered to the substrate and have grouted joints. On the exterior, they are applied as a rainscreen. Given the number of units to install, a very simple connection strategy was required. Black coated aluminum straps with projecting eyelets are glued to the back rim of the hollowed out shapes. Mounting screws can then be inserted through the open joints into aluminum rails. These rails run parallel to each other, just to one side of the

Figure 6.83 Trumpf Campus Restaurant, Ditzingen, Germany (2008). Barkow Leibinger, Architects, Berlin. Kite shaped terra cotta tiles by NBK Keramik. Applied to the exterior as a ventilated rainscreen. Similar, polychromatic tiles adhered to interior surfaces. Photo: Christian Richters.

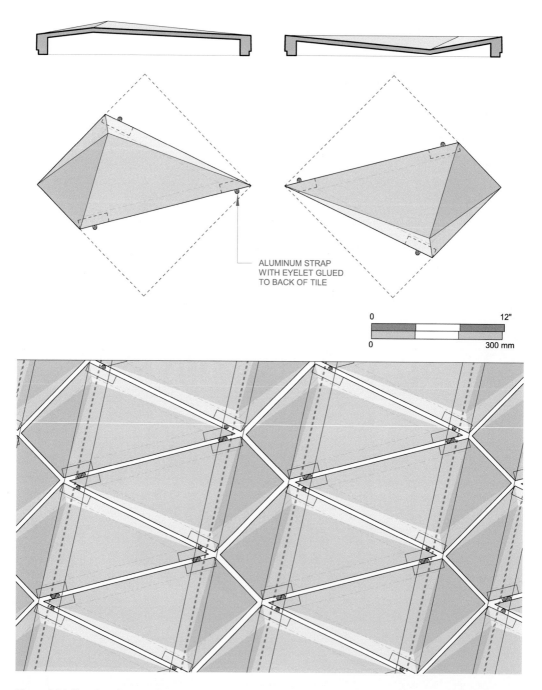

ALUMINUM STRAP
WITH EYELET GLUED
TO BACK OF TILE

Figure 6.84 Elevation diagram. Campus Restaurant, Ditzingen. Ram pressed tiles with asymmetrical convex and concave profiles, attached to aluminum "T" profiles with screws driven through the open joints. From drawings by Barkow Leibinger, Architects.

points of intersection of four tiles. This provides three points of connection for each unit in an efficient, recurring pattern (Figures 6.84 and 6.85).

The Asian Art Museum of San Francisco relocated in 2003 to the former city library. The 1917 Beaux Arts building was renovated as a museum by

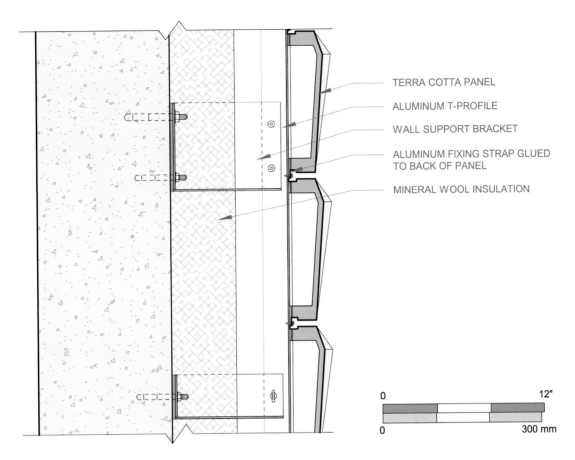

TERRA COTTA PANEL

ALUMINUM T-PROFILE

WALL SUPPORT BRACKET

ALUMINUM FIXING STRAP GLUED
TO BACK OF PANEL

MINERAL WOOL INSULATION

0   12"

0   300 mm

Figure 6.85 Vertical section through convex wall tiles. Mounted on aluminum rails in front an insulated wall cavity. Campus Restaurant, Ditzingen. From drawings by Barkow Leibinger, Architects.

Italian architect Gae Aulenti. In 2016, the museum announced plans for an addition to provide a special exhibition space and other visitor attractions not available in the original building. Undertaking the design, wHY Architecture had to address two strong themes: the historic character of the granite clad civic structure and the modern spirit of the institution in its renovated space. The addition is constructed on top of a service wing at the rear of the building; a location that permitted large openings for views outward to life on the street and inward to the exhibition pavilion (Figure 6.86).

Working with the Historic Preservation Commission and Planning Department, the architects proposed a uniquely textured façade that would respond to the rusticated granite blocks at the base of the historic building. Significant areas of glazing were a given, but shaped metal panels were considered too great a departure from the context. Terra cotta has a long tradition of interacting with granite, both in terms of the forms and the finishes. Large-scale rectangular panels were developed, with a new, angular shape. The large glazed opening on the gallery level has the same

Figure 6.86 Addition to the Asian Art Museum of San Francisco, California, by wHY Architecture (2020). Ram pressed terra cotta panels create a contemporary, rusticated surface. The motif is picked up in the shapes of the glazing overlooking the street. Boston Valley Terra Cotta. Photo: Jenny Young.

geometry, projecting outward in the manner of San Francisco oriel windows (Figure 6.87).

Boston Valley Terra Cotta prototyped the panel forms by hand pressing. After review and approval, they built more technically demanding ram press molds for volume production of the repeating units. Edge conditions requiring a slightly different mold and fewer repetitions were produced as solid, slip cast units. The panels were designed with shiplap joints that allowed them to be installed with a standard track and clip system.

The architects asked for a finish that would resonate with the granite blocks of the main building while expressing a new and different material. Working with Boston Valley, they selected a light gray matte glaze, with a spattered application of dark at a higher gloss. The combination has more sheen than granite, creating a further distinction between the two materials.

One of the most extraordinary aesthetic accomplishments in contemporary terra cotta is the extension of the Center for Asian Art at the Ringling Museum in Sarasota, Florida (2016). Designed by Machado Silvetti in collaboration with Boston Valley, it captures the volumetric traditions of the material, the capacity for sculpted surfaces and rich glaze effects, all in a modern form.

The Ringling Museum of Art opened in 1931, with long, thin gallery wings forming three sides of a courtyard filled with sculpture. Additional facilities were added to the western ends of the "U" opening toward Sarasota Bay. One of these wings houses a study center and a significant collection of Asian art. The attached pavilion provides a visual anchor for the complex and a shaded, on grade, entry point from extensive grounds along the waterfront.

The new construction contains a special exhibition space, lifted up to align with existing galleries on the second level. A lecture theatre connects to the study center on the top floor. This simple, compact program allowed the architects to focus on the innovative façades. First sketches show tessellated surfaces with openings for windows within rather than between the units. Through a series of geometric operations, a square grid was transformed into nested, small and large squares, rotated eight degrees from vertical. Each of the elements has a warped surface and crisp edges inspired by roof tiles in Asian architecture. The larger squares are occasionally punched for windows. The green glaze derives from the jade collection held within and the lush colors of the surrounding garden. Double firing allowed the color to pool in the hollows, adding another surface effect to those created by shadow and reflection (Figure 6.88).

The repeating tiles in the field were ram pressed. The larger units are 23–5/8" x 24-1/2" x 5-1/2" (58 x 61 x 12.8 cm) with a nine square grid of hollows formed at the back. The complex shape demonstrates the plasticity of clay and the deep draw available at the press.

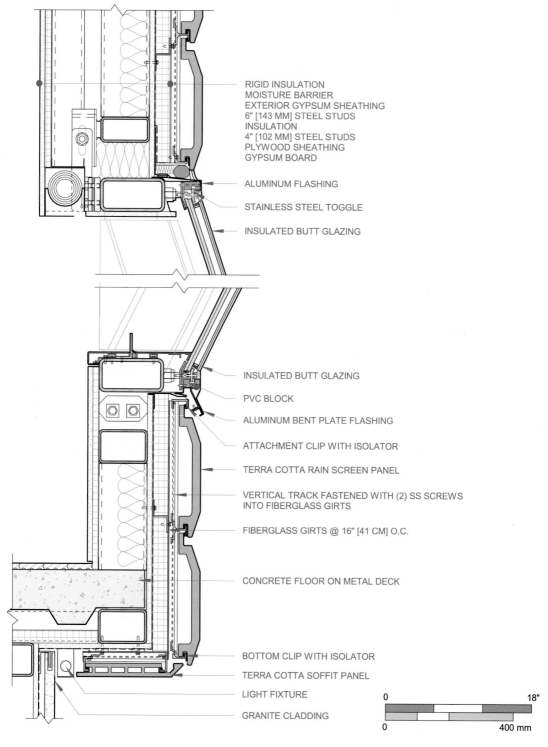

RIGID INSULATION
MOISTURE BARRIER
EXTERIOR GYPSUM SHEATHING
6" [143 MM] STEEL STUDS
INSULATION
4" [102 MM] STEEL STUDS
PLYWOOD SHEATHING
GYPSUM BOARD

ALUMINUM FLASHING

STAINLESS STEEL TOGGLE

INSULATED BUTT GLAZING

INSULATED BUTT GLAZING

PVC BLOCK

ALUMINUM BENT PLATE FLASHING

ATTACHMENT CLIP WITH ISOLATOR

TERRA COTTA RAIN SCREEN PANEL

VERTICAL TRACK FASTENED WITH (2) SS SCREWS
INTO FIBERGLASS GIRTS

FIBERGLASS GIRTS @ 16" [41 CM] O.C.

CONCRETE FLOOR ON METAL DECK

BOTTOM CLIP WITH ISOLATOR

TERRA COTTA SOFFIT PANEL

LIGHT FIXTURE

GRANITE CLADDING

0		18"

0		400 mm

Figure 6.87 Asian Art Museum, San Francisco. Vertical section at the head and sill of the faceted, oriel window on the street façade. Ram pressed, shiplap panels by Boston Valley Terra Cotta. From drawings by wHY Architecture, Culver City, California.

Figure 6.88 Center for Asian Art at the Ringling Museum, Sarasota, Florida (2016). Architects: Machado Silvetti, Boston, Massachusetts. Deep, ram pressed units, alternating as large and small squares on a rotated grid. Boston Valley Terra Cotta. Photo: © Anton Grassi/Esto.

After pressing, grooves were cut along the bearing edge of each unit to complete the shiplap joints used for a standard Boston Valley rainscreen attachment system. The system had the advantage of prior approval for use in a hurricane zone.[26] Short lengths of supporting rail must follow the eight-degree rotation of the nested squares. This complex pattern is fastened to a continuous CMU backup wall. Windows occur in clusters on the façade. In those zones, the CMU is replaced with horizontal tube steel that supports the rail system at regular intervals. The punched openings in the terra cotta are counter rotated back to normal so that the glazing units fit between the steel (Figure 6.89).

The two US based firms, Gladding McBean and Boston Valley, actively produce hand pressed terra cotta units. The majority are replacement parts for structures of the late nineteenth and early twentieth centuries. The standards of historic preservation have risen, increasing the demand for in-kind replacement, as opposed to substitute materials. This has returned tremendous benefits to the industry, advancing the skills of those who form and finish classic components. Volumetric hand pressed units find their way into new construction where highly customized pieces are used in combination with repetitive parts formed by other means. Such a project is The Fitzroy in New York, which is included as a case study in Chapter 9.

The versatility of slip casting has a role in contemporary work as the means to produce unique, three-dimensional parts that are either too complex or too few in number to be made by other methods. This supplemental role was described in relation to the largely ram pressed projects illustrated in the previous section.

A true distinction of slip casting is the ability to make closed forms that are hollow, following the model of sanitary ware (toilets) that we so easily take for granted. The internal void might be inaccessible, or simply too small for the hands of the worker pressing clay into an open backed form. The roof crest of Chapter 3 combines solid and hollow portions in a single form that could not be achieved in any other way (see Figure 3.13).

Slip casts remain a craft product, accurately reproducing complex forms for clients who prefer a single, integrated piece. They represent one true potential of terra cotta. Both Palagio Engineering and Boston Valley prototyped elements for a project at London Wall Gate (Figure 6.90). At large scale, and in repetitive quantity, these traditional methods have to compete with all of the sophisticated new ways to cut and fit extruded pieces to make a composite form, often using glued joints rather than monolithic construction.

At a slightly smaller scale, slip cast parts are found where terra cotta accents are incorporated into glazed walls and screens. Palagio made mullion

[26] Gulling, p. 341.

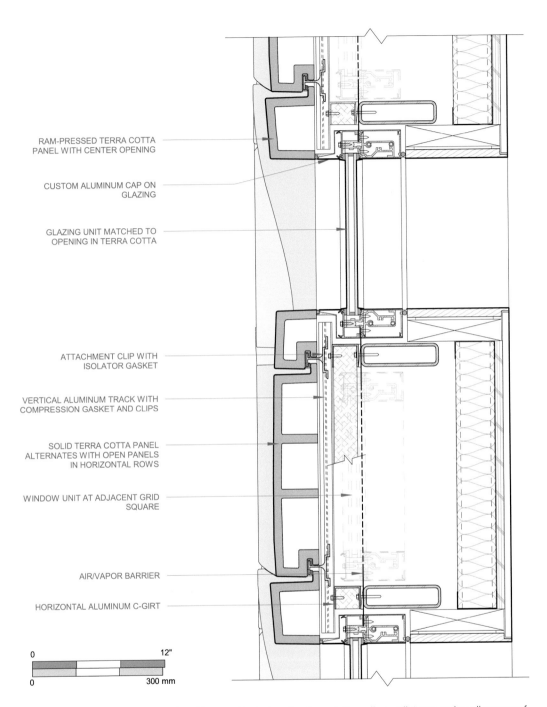

RAM-PRESSED TERRA COTTA
PANEL WITH CENTER OPENING

CUSTOM ALUMINUM CAP ON
GLAZING

GLAZING UNIT MATCHED TO
OPENING IN TERRA COTTA

ATTACHMENT CLIP WITH
ISOLATOR GASKET

VERTICAL ALUMINUM TRACK WITH
COMPRESSION GASKET AND CLIPS

SOLID TERRA COTTA PANEL
ALTERNATES WITH OPEN PANELS
IN HORIZONTAL ROWS

WINDOW UNIT AT ADJACENT GRID
SQUARE

AIR/VAPOR BARRIER

HORIZONTAL ALUMINUM C-GIRT

0   12"

0   300 mm

Figure 6.89 Center for Asian Art, Ringling Museum. Vertical section through the gallery wall. Large and small squares of terra cotta alternate in rows that are framed with hollow steel tubes. Large squares have central voids that open to the window units. Machado Silvetti, architects. From shop drawings by Boston Valley Terra Cotta.

covers for the Francis Crick Institute in London this way (HOK Architects, 2017). The projecting fins for Eric Parry's Holburne Museum in Bath were also slip cast by Darwen Terra Cotta and Faience (Figure 6.91), finished in a splattered glaze that resonates with the garden setting (Figure 6.92).

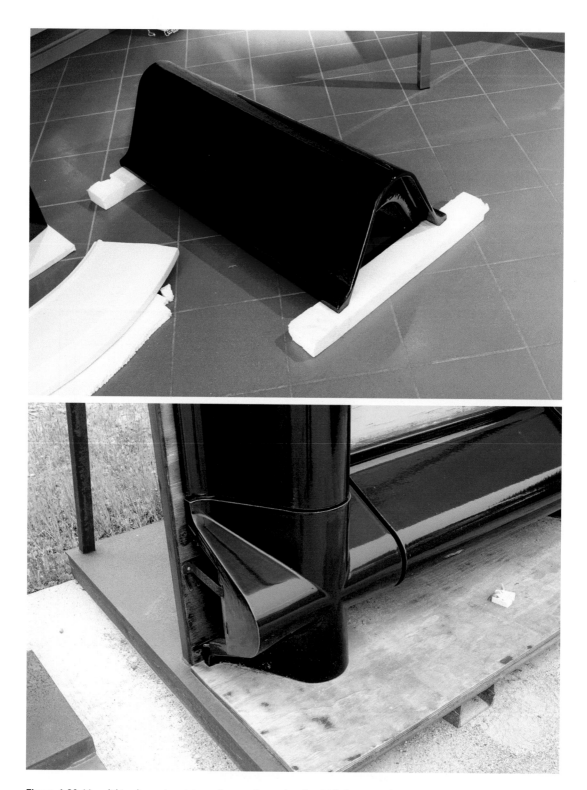

Figure 6.90 Monolithic, slip cast prototypes for a project at London Wall Place, London, UK. Make, architects. Above: Single unit at Palagio Engineering. Below: Façade bay mockup at Boston Valley Terra Cotta. Photos: Donald Corner.

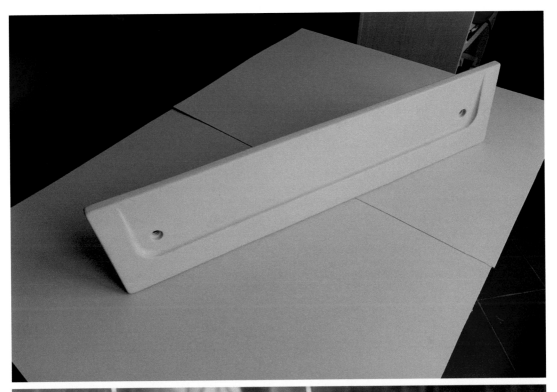

Figure 6.91 Above: Slip cast mullion cover for the Francis Crick Institute, London, UK (2017). HOK Architects. Palagio Engineering. Photo: Donald Corner. Below: Slip cast fins with a splattered glaze prepared for firing. Holburne Museum, Bath, UK (2001). Eric Parry Architects. Darwen Terra Cotta and Faience, Blackburn, Lancashire, UK. Photo: Grant Smith.

Figure 6.92 Holburne Museum, Bath, UK (2001). Eric Parry Architects. Detail of slip cast terra cotta fins with splattered glaze. Photo: John Rowell.

CHAPTER 7

Intelligent Variation

In recent decades, the concept of "mass customization" has been used to describe various paths to consumer satisfaction through affordable, quality products.[1] Since it was coined, the term has become broadly inclusive in application. In the most general sense, it describes the use of new technologies, often information technologies, to combine client driven customization with the efficiencies and cost benefits of mass production. In different modes of mass customization, the client's influence can be brought to bear on design, fabrication, post-production or final assembly.

This combination of benefits can be approached from either direction. Bespoke, or engineered-to-order, products can be moved toward production strategies that are more efficient. Conversely, methods of mass production can be opened up to respond to inputs from the consumer, beginning with something as simple as a choice of colors. The ultimate goal is to provide intelligent variations that respond to particular needs of the client, while they respect the acquired knowledge of efficient production. Owing to its literal plasticity, terra cotta is a medium that adapts perfectly to this search for creative balance. Terra cotta has been in continuous production from pre-industrial contexts to the present, from hand crafting of artifacts to the highly automated production of repetitive parts.

Central to the notion of mass customization is that the client be admitted to a role as co-designer of the product, while the fabrication process remains flexible, stable and reasonably close to the efficiencies of mass production. The cost of customization should not require the shift to a higher market segment.[2]

Over the history of terra cotta, economic advantages have been greatly dependent on context. During the golden age of hand packed façade components, terra cotta was competing with stone. The repetitive use of molds was profoundly efficient compared to the hand carving of each piece.

[1] S.M. Davis.
[2] Piller, 2004, as quoted in Haug et al., p. 4.

DOI: 10.4324/9780429057915-8

A great deal of customized design could be added without threatening the overall savings. At the mass production end of the spectrum, present-day terra cotta competes with materials like glass fiber reinforced concrete (gfrc) to produce durable, flat and lightweight rainscreen cladding panels. The opportunities for customization are limited to color and surface texture within a prescribed range of thicknesses and attachment techniques.

The contemporary components and systems featured in this book derive from the upper end of the present market. They represent terra cotta as a high value material, operating in a cost niche that leaves room for a fascinating range of custom design variations. It would be difficult to argue that they are all economical, although one can see how they are influenced by rationalized means of production. While the terra cotta pieces are not inexpensive on their own, they have an appropriate place in high-value, long-term building projects. Success is defined as added value that remains at least proportional to added price.

Of the many paths to mass customization, there are two of particular interest in relation to terra cotta. First, there is the use of templates, either mechanical or digital, that allow a wide range of design options to be realized within a coherent system of efficient production. Second is the mixing and matching of both general and particular parts to create custom, or semicustom final assemblies. Terra cotta has been worked in both of these ways throughout its history. It is in many respects an ideal medium.

At the Etruscan archaeological site of Poggio Civitate (Murlo, Siena), the "Orientalizing Complex," from the seventh century BCE, was roofed using flat terra cotta tiles with upturned edges (tegula) and convex tiles (imbrex) to cover the joints. The basic tiles repeat across the broad expanse of the roof, but the lowest course is turned up to form the upper portion of a terra cotta cornice. Rainwater is released through feline spouts and the cover tiles are terminated with antefixes in the shape of female heads. The later "Archaic Building," after 580 BCE, featured the previously mentioned Murlo Cowboy (see Figure 3.1). In both cases, a distinctive roof was achieved with a limited number of custom pieces.

In the Shaanxi Province of China, the famous Terra Cotta Army guards the tomb of First Emperor Qin Shihuangdi. Beginning as early as 221 BCE, thousands of life-sized soldiers were produced using interchangeable, previously molded parts: heads, torsos, tunics, arms, hands, legs, feet and plinths. Heads were finished by hand, applying different ears and eyebrows. The parts were joined before firing, then lacquered, making a vast array of different outcomes within a highly organized system of production.[3]

In common buildings of stone construction, the detailed and carefully cut pieces are reserved for the bonded corners (quoins), cornices

[3] Ledderose, pp. 68–73.

and window surrounds. The surfaces between can be filled in using rough cut ashlar blocks, or even rubble stone, readily placed by the mason. As molded terra cotta took the place of the carved stone, efficiencies of mass production could be applied to what had been the most labor-intensive parts of the overall stone assembly (Figure 7.1).

Continuous terra cotta elements, integrated into solid masonry construction, represent a relatively low threshold of customization. String courses, water tables, cornices and parapet caps may be custom designed (made-to-order) for a given project, or they might simply be chosen from previously designed pieces (made-to-stock) held in readiness by the producer. Making pieces to stock allows a facility to remain active though the rise and fall of custom orders. The marketing success of Eleanor Coade, introduced in Chapter 2, incorporated stock elements that architects could adopt when preparing a building elevation. "The 1784 catalogue contained no fewer than 788 designs. Often the pieces could be customized: a goddess's face given different headdresses, columns and capitals mixed and matched, chimney pieces assembled by assortment."[4]

[4] Stanford, pp. 10–11.

Figure 7.1 Ornamental boundary elements applied to a rough masonry wall. Left: Casa Longega (fifteenth century), Conegliano (Treviso), Italy. Right: House of Giovanni Boniforte da Concorezzo (1455), Mantova, Italy. Photos: Donald Corner.

The California company Gladding McBean developed their own forms of mass customization to extend the market reach of their products.

In 1927, the Tropico Plant published a catalogue for stock architectural terra cotta "developed for the purpose of facilitating the use of Terra Cotta on the smaller buildings where lowest costs and shortest deliveries consistent with good manufacturing are imperative." Both the Glendale and Lincoln plants carried stock molds for urns, ornaments, cartouches, pilasters, ashlar, cresting, panels, cornices, and architraves. Available in cream enamel finish, the stock items could be delivered within a week to ten days of an order. For an additional 25%, the company would produce pieces in polychrome. Examples of stock terra cotta produced by these two plants appeared on the facades of automobile dealerships, schools, banks, hotels and businesses throughout the West.[5]

Terra cotta has been used to create areas of intense detail on a wall or façade since the earliest days: sculptural panels, rondels and lunettes. Best known, perhaps, are those of the Della Robbias in Florence. The concept of focused detail was applied to early American high-rises, using terra cotta to differentiate spandrel panels from the vertical, load bearing walls and piers of brick. The numerous, repetitive bays of these buildings brought down the unit cost of molded, ornamental embellishments. With the maturity of the structural frame, in iron, steel and concrete, terra cotta took over the entire, non-bearing façade. This brought production efficiencies to both the figure and the field.

The transition to "machine made" terra cotta in the 1930s further rewarded the use of selective detail. Large areas of the building elevation could be efficiently covered with elements that were essentially industrial products. The tiles became flatter with relatively gentle custom textures on the face of broad panels. Variety from building to building could be expanded through the creative use of saturated glazes that were reaching new levels of chemical sophistication in the same period. There was a perfect match between the forms arising from the process and the sensibilities of the Art Deco style. Traditional, molded elements brought strong contrasts to the smooth lines and surfaces particularly where vertical elements met the sky (Figure 7.2).

Contemporary production of terra cotta is dominated by the extrusion process, and the agent of mass customization is the extrusion die. For a given project it is possible to select profiles from a library of existing dies held by the producer. It is also possible to design for new, custom dies.

[5] Kurutz, pp. 123–4.

Figure 7.2 Selective use of relief combined ceramic veneer panels. Above: Parapet of the Breuner Building, Oakland, California (1931). Designed by Albert Roller. Below: Corner tower of the Wiltern Theatre Building, Los Angeles, California (1931). Designed by G. Albert Lansburg. Photos: Donald Corner.

Although they are expensive, each die constitutes a very small part of the overall production machinery, while all of the rest remains unchanged. The cost of a custom die is further reduced by cutting them with automated milling machines, or water jet rigs, that are driven directly from digital design models.

For most of building history, the opportunity to customize the layout and orientation of masonry units has been facilitated by the universal mortar joint between them. Brick, stone and terra cotta can be freely mixed and matched within this system. The only limitation is the ability of traditional mortars, or modern sealants, to adhere to the materials on opposing sides of the joint. They must not tear loose due to long-term movements created by thermal or moisture effects. The open joints of a contemporary wall eliminate these constraints. The affinity of terra cotta for rainscreen applications means that it can be juxtaposed with an extraordinary range of materials that it need not ever touch.

Within a field of terra cotta units, shiplap profiles enhance the rainscreen performance of horizontal joints. Every major producer offers a variation on the basic detail promoted by Thomas Herzog (see Figure 5.4). This deceptively simple configuration sheds water, facilitates a standardized system of attachment, and unlocks a universe of custom design options. The requisite forms and dimensions at the joint can be developed on components that have been extruded, ram pressed or slip formed. The majority of horizontal systems presented in Chapter 6 rely on this approach: virtually all of the simple planks, the prismatic elements of the Asian Art Museum in San Francisco (see Figure 6.87), and the warping volumes of the Ringling Museum of Art in Florida (see Figure 6.89).

Vertically oriented terra cotta units are supported from below, most often using a simple end cut on the extruded profile. Standardized means of attachment are facilitated by a limited range of wall thicknesses between the back of the extrusion and the voids of the internal cores. Great variations of the outer face can be generated by changing the die. Furthermore, the variations can be doubled in number by rotating an asymmetrical piece end for end with little or no change required in the systems of attachment (see Figure 6.40). At the Kirkland Museum in Colorado, variations are created using rectangular baguettes that can be attached with either the broad or narrow edges against the building (see Figure 6.44).

The Eddie and Lew Wasserman Building (2014), on the campus at UCLA, uses a simple exchange of extruded components to achieve a transition across the façade, from open to closed. Designed by Richard Meier and Partners, the building is organized by a terra cotta clad wall plane, facing onto an arrival garden. Public spaces occur in front of this plane with clinic spaces behind. The wall faces south, opened by strips of vision windows and clerestories, with baguette screens (Figure 7.3). The horizontals of the baguettes are carefully aligned with profiles on the adjacent rainscreen cladding. The ribbed profiles systematically die out to complete the transition to a solid wall. Thus, a large-scale, custom expression is achieved with a relatively small set of mass produced parts, facilitated by a standardized support system (Figure 7.4).

A larger inventory of custom shapes can be created by post-processing basic extrusions. Boston Valley Terra Cotta has completed several examples in New York City. The garden wall of a townhouse on the Upper East Side is clad with terra cotta planks extruded as a double stemmed tee (Michael K. Chen, Architects, 2014). The stems, which project outward, are variously trimmed to produce an undulating pattern of form and shadow. The spandrel panels of an office building at 512 West 22nd Street are clad with extruded foil shapes for a bold, horizontal shadow pattern (COOK-FOX Architects, 2018). The green clay was gently reshaped so that the foils twist open as they wrap around the curving corners of the building.

Figure 7.3 Façade detail from the Edie and Lew Wasserman Building at the University of California, Los Angeles (2001–14). Designed by Richard Meier and Partners, Architects. Photo: Donald Corner.

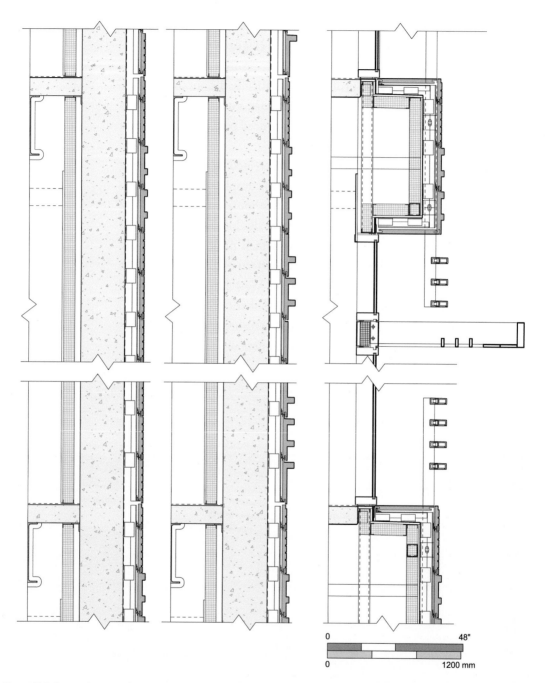

Figure 7.4 Successive vertical sections at the Wasserman Building, UCLA. Hand set terra cotta planks are exchanged with custom profiles to create the transition seen in Figure 7.3. Ribbed panels are aligned with baguette screens at the window openings. From drawings by Richard Meier and Partners, Architects.

The cutting and reshaping for these two projects was executed within the craft traditions of terra cotta. Their success motivated the development of tools to do similar manipulations in higher volume, under digital control. Evolving out of their Architectural Ceramic Assemblies Workshop

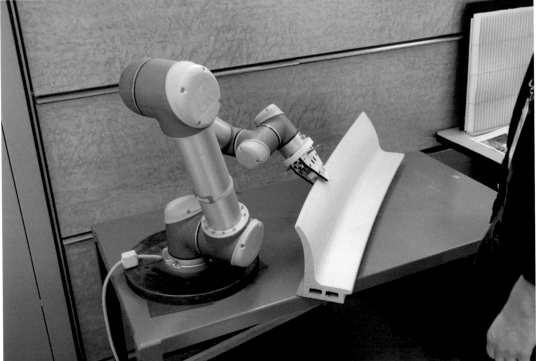

Figure 7.5 Concepts of customization on display at the 2019 Architectural Ceramic Assemblies Workshop. Above: Multi-axis wire cutting of extruded profiles by SOM, Architects. Below: Reshaping green terra cotta with a robotic arm, by Pelli-Clarke-Pelli and Studios Architecture. Technical support by Boston Valley Terra Cotta. Photos: Donald Corner.

(ACAW), Boston Valley Terra Cotta developed a multi-axis wire cutter that can slice through a clay extrusion in virtually any sequence of angles as the clay moves past the wire. ACAW participant teams lead by SOM, Architects, experimented with this capability during the 2019 and 2020 sessions (Figure 7.5, top image). A team from Morphosis Architects began to experiment with the curving and twisting of extruded sections during ACAW 2017.[6] The fruit of this collaboration with Boston Valley will debut at full scale with completion of the Orange County Museum of Art (2021–2). In 2019, a team from Pelli-Clarke-Pelli and Studio Architecture projected the future use of a robotic arm to reshape green clay through successive passes (Figure 7.5, bottom image).

Digital design and process control will continue to bring economies of rational production to a universe of complex forms that can be realized in clay.

[6] Garófalo and Kahn, 2017, pp. 48–59.

CHAPTER 8

Design Imperatives

What do the buildings and details presented in this book reveal about the essence of terra cotta? What are the imperatives for its successful use in design? Asking this latter question about a material is fundamental to the development of architecture, but it does not always lead to simple answers. Terra cotta is a material of incomparable range that has been skillfully adapted to the needs and goals of an endless stream of very different builders.

There are two major epochs in the recent history of terra cotta that are important to review: the 1870s through 1930s, and the 1980s through to the present. Both are periods of tremendous growth and change in the volume of product applied and in the driving ideas that called for the material.

The introduction, or revival, of a construction material often begins with mimicry. Forms in the new follow familiar traditions of the original. So cast iron façades imitated cut stone before finding their own expression. Prior to the 1870s, terra cotta was trapped in this phase. Architects influenced by the earlier writings of Pugin and Ruskin considered any ornament that did not result from hand working of authentic materials to be dishonest, or fake. Pamela Simpson chronicles this period in her book, *Cheap, Quick, and Easy: Imitative Architectural Materials, 1870–1930*. She includes evocative quotes. In London, the *Building News* referred to terra cotta on the Victoria and Albert Museum (1864–7) as "sham columns in a casing of crockery."[1] When James Taylor was brought from England to build up the Chicago Terra Cotta Company, he described the previous production of the firm as "a clay imitation of an iron imitation of cut stone."[2] Taylor went on to become the "Father of Terracotta in America," working with innovative architects to use the material on full building façades, rather than just the occasional, imitative insert.[3]

[1] Simpson, p. 130, note 54.
[2] Simpson, p. 130, note 56.
[3] Simpson, p. 130, note 58.

DOI: 10.4324/9780429057915-9

Pamela Simpson outlines the "vast industrial and social changes" that occurred in this time interval.[4] These included the emergence of new commercial building types, particularly skyscrapers, in which terra cotta played an essential role. Across the period there were waves of new ideas and great changes in architectural style. At the mid-point, Louis Sullivan coined the phrase, "form follows function," in his seminal article on tall buildings.[5] Terra cotta was the material through which these new ideas were realized. The epoch began with intricate replicas of classical forms and ended with broad expanses of extruded, ceramic veneer, the machine made terra cotta. The essence of terra cotta was its unique ability to facilitate all of this change. Initially decried as imposters, landmarks from early in the period have become revered, as by Virginia Guest Ferriday in *The Last of the Handmade Buildings*. When the economics of industrial production streamlined the forms, the inventive terra cotta industry responded by adding brilliant color.

In the second epoch, the 1980s to the present, terra cotta was "rediscovered" for strict, rational applications, but has subsequently moved back toward the volumetric exuberance of earlier times. After World War II, the continuing growth of frame and infill buildings pushed the envelope inexorably toward lighter, thinner, flatter skins, dominated by metal and glass. Brick cladding still remains a practical, durable and cost-effective choice, but the overwhelming majority of applications are single wythe, anchored veneers. In this use, brick poses the same intellectual challenge that it did during the birth of the skyscraper. We respond positively to its warmth and character, but how should we express the fact that it does not support the building? Although rainscreen wall design was introduced a decade earlier, by the 1980s it became the unifying theory. Through the ventilating joint, air replaced mortar as the modern, universal interface between parts. It allowed free composition in a range of materials that could change in size, shape and orientation. At the same time, the open joint declares unequivocally that cladding is not the primary structure. Terra cotta was given new life in the rainscreen wall. Through the developmental work of Thomas Herzog and Renzo Piano, the extruded plank became the new brick. It is lighter and larger for ease of installation, yet it retains the properties we admire in masonry products: scale, texture, color and ornament.

The tight, technical rainscreen walls of Thomas Herzog became very popular with architects and suppliers. Palagio Engineering, for example, was a high volume producer of floor tiles that then migrated through several steps to the extruded, shiplap plank. Flat, modular planks with natural finishes are a pure form in principle, but they do not capture the full potential of terra cotta. An obsession with flatness opens the door to

---

[4] Simpson, p. 136.
[5] Sullivan, 1896.

substitute materials, like cementitious composites that are less expensive, but lack the same depth of character.

The flat back of extruded products facilitates their passage through the stages of production, but, as we have seen, that need not constrain the form of the front face. Natural clay body colors are beautiful, but terra cotta really distinguishes itself from the competition with the application of glaze. The works of Sauerbruch Hutton demonstrate that simple shapes can produce dynamic results with the creative use of color. Over the last decade of this epoch, pioneering firms like Palagio Engineering have moved away from high volume production of standard shapes to high quality, custom work, imbued with all of the qualities the material can deliver.

The resurgence of terra cotta has been supported by another trend. Since the 1980s, there has been a shift in historic preservation toward greater use of authentic materials rather than substitutes.[6] Thus, landmark buildings of the earlier period have spawned aspects of the current revival. Efficient production of replacement parts for classic buildings has reinvigorated the skills and techniques available to new construction. A prime example is the Ringling Museum addition presented in Chapter 6. Machado Silvetti, Architects and Boston Valley Terra Cotta developed a thoroughly modern integration of sculptural, polychrome terra cotta and the shiplap, rain-screen technology of the present day.

As the perfect, plastic medium, terra cotta has been given its form by the architectural ideas of each epoch. What are the concepts that will shape the material going forward? Foremost among them must be environmental goals: building efficiency, longevity and reduced carbon impact.

To optimize energy performance, architects have rediscovered the importance of solid walls, moving on from a period in which buildings have been consistently overglazed. Terra cotta has always been able to bring meaning and delight to solid walls. Ecological architects have a new appreciation for durability and low maintenance. Intricate, all glass double façades have fallen out of favor because they are so difficult and expensive to maintain. Terra cotta has been the opposite.

The critical, environmental measure of present and future building is "whole life carbon," accounting for the emission of greenhouse gasses due to production, construction, operation, disassembly and re-use. As we have systematically improved the operating efficiency of our structures, the emphasis has shifted to their "embodied carbon" content. Despite the obvious energy inputs, fired ceramic products have favorable ratings when compared to other, highly industrialized commodities, particularly aluminum. To realize its potential, terra cotta must reduce its dependency on aluminum to complete construction assemblies.

[6] Henry et al., p. 681.

Figure 8.1 Chiesa della Misericordia, Terranuova Bracciolini (Arezzo), Italy (2015). Archea Associati, Architects. Exterior cladding consists of only two types of hexagonal façade tiles by Tagina Ceramiche d'Arte: flat and prismatic. Photo: Donald Corner.

Figure 8.2 Chiesa della Misericordia. Surface articulation generated by varying the orientation of the asymmetrical, prismatic façade tiles and interspersing flat tiles of the same base dimension. Photo: Donald Corner.

The calculus of carbon impact for a given material is most readily improved by dividing it over a longer service life, again favoring terra cotta. Open jointed rainscreen systems lend themselves to recovery, re-cutting and re-use on a new building. This intrinsic advantage could, paradoxically, drive the industry back to systems of regular modules, and away from the fully customized dimensions of recent years.

Alessandro Piazza, export manager at Palagio Engineering, describes the effective use of terra cotta in terms of Lego sets.[7] Traditionally, young builders exercised their ingenuity by making wonderful things from combinations of a few typical bricks. Modern sets try to anticipate the desires of the builders by providing specialty parts for every conceivable circumstance. While this may be nearly possible, he argues that it does not necessarily unlock the same level of creativity (Figures 8.1 and 8.2).

The ideas that drive the use of architectural terra cotta circle back on themselves over time, even as the technologies continue to evolve. What remains unchanged is the incomparable ability of this unique material to be the agent of imagination.

[7] Palagio Engineering, 2019.

CHAPTER 9

Case Studies

The case studies of this final chapter have been chosen to underscore major themes of the book. Contemporary applications of terra cotta are manifestations of history; traditions of the material itself and innovations within the building technologies most closely associated with its use. The comparative discussion in these studies places the selected projects within the evolving story of the material.

The Fitzroy, in New York, features a highly customized, hand pressed façade that invites comparison with the golden age of terra cotta in the 1920s and 1930s.

Potsdamer Platz, in Berlin, captured the interest of architects around the world and was a catalyst for the most recent revival of terra cotta as rainscreen cladding.

Central Saint Giles, in London, rekindled the tradition of intense color in glazed terra cotta, and that of depth in the façade, achieved through layering of simple extruded sections.

125 Deansgate, Manchester (UK), uses slender, rationalized components to bring depth and warmth to the largely transparent façades of a contemporary office building. It invites comparison to the revolutionary "skyscrapers" of Chicago, in which terra cotta played such an important role.

Tykeson Hall, at the University of Oregon, reunites terra cotta and brick. As a strictly disciplined, almost "off the shelf" application, the project demonstrates the richness of the material even without the customization of other cases.

DOI: 10.4324/9780429057915-10

The Fitzroy, New York, NY (2019)
Roman and Williams: Buildings and Interiors

The Fitzroy is a boutique condominium on West 24th Street, West Chelsea, adjacent to the famous High Line of New York (Figure 9.1.1). It was designed by Roman and Williams in collaboration with JDS Development Group. The terra cotta façade was inspired by Art Deco buildings in the city, but not the immediate neighborhood, which has many new buildings of limited detail. Founders Robin Standefer and Stephen Alesch were alert

Figure 9.1.1 Central portion of the elevation of The Fitzroy, 514 West 24th Street, New York, NY (2019). Designed by Roman and Williams for the JDS Development Group. Façade cladding by Boston Valley Terra Cotta, Orchard Park, New York. Photo: Stephen Alesch.

DOI: 10.4324/9780429057915-11

for an opportunity to do a highly developed, craft-oriented building. They were approached by JDS, who by this time had completed several projects in buildings by Ralph Walker, the preeminent architect of Art Deco skyscrapers in the 1920s.

Massing of The Fitzroy derived from the "dormer law" which allows shaped volumes to return to the street face within the characteristic setbacks of New York zoning.[1] These regulations were the catalyst for stepped terraces, buttresses and pilasters that were ultimately developed in the terra cotta cladding.[2] The zoning requirements were captured with an early concept drawing, done in pencil by Stephen Alesch (Figure 9.1.2). Supplied with a catalog of traditional extruded profiles by Boston Valley Terra Cotta, he then developed details for the vertical and horizontal features of the façade. After refinement, in both hand and digital media, the scheme was committed to high-end digital renderings for promotion and sales.

[1] Zoning Resolution of the City of New York.
[2] Alesch.

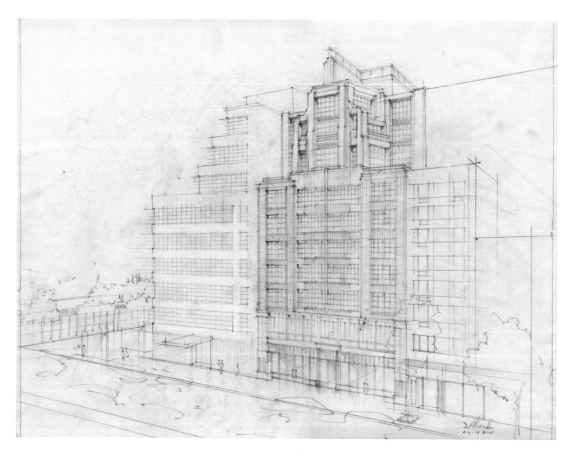

Figure 9.1.2 Early study drawings of the massing and elevation development for The Fitzroy. Responds to the setback provisions of New York zoning. Drawing by Stephen Alesch.

Stephen Alesch continued to draw over the renderings to adjust the forms of the elements and the transitions between them (Figure 9.1.3). Having started with a genuine effort to use "stock" parts, the project evolved into 5,600 terra cotta blocks with 500 unique types. Working back and forth between CAD and Rhino, digital models were developed for each piece.[3] Boston Valley refined the pieces for constructability and Buro Happold designed a suspended shelf angle and clip system to support them (Figure 9.1.4). This system was applied regularly across the façade. The layout of wall studs behind the barriers and sheathing was coordinated with the desired location of the clips. This greatly simplified the installation

[3] Devine.

Figure 9.1.3 Design refinement of the terra cotta cladding for The Fitzroy. Pencil sketch over a color printout from a rendered, 3-D digital model of the project. Drawing by Stephen Alesch.

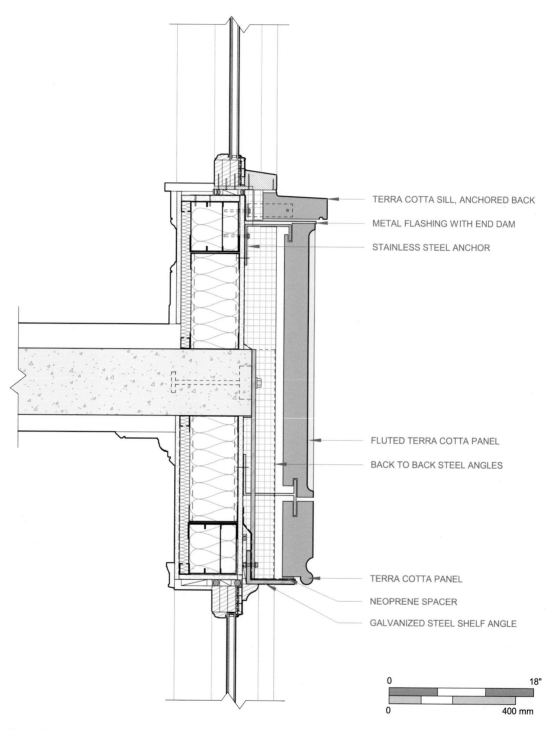

TERRA COTTA SILL, ANCHORED BACK

METAL FLASHING WITH END DAM

STAINLESS STEEL ANCHOR

FLUTED TERRA COTTA PANEL

BACK TO BACK STEEL ANGLES

TERRA COTTA PANEL

NEOPRENE SPACER

GALVANIZED STEEL SHELF ANGLE

Figure 9.1.4 Vertical section taken at a typical spandrel panel in the central façade of The Fitzroy. Terra cotta supported on shelf angles, suspended from the slab edge. Anchor locations coordinated with layout of the steel stud infill wall. Design Team: Roman and Williams, IBI Group – Gruzen Samton, and Buro Happold Consulting Engineers.

process. The thick wall section was only possible because of the developer's willingness to sacrifice some interior floor space for a deep, articulated façade.

Elements of the façade were dry fit on the floor at Boston Valley before shipping.[4] A vertical mockup was constructed outside using plastic shims where the uniform mortar beds would ultimately occur. In addition to fit, this allowed the layered green glaze to be reviewed and approved in various angles of daylight. On site, each of the blocks was hand set by masons over a period that extended from August 2017 to August 2018. The copper clad windows were installed from the inside to separate the trades.

At the top of the building, a full repertoire of terra cotta shapes and volumes came into play. The parapet of the upper roof terrace shows contemporary, extruded panels suspended as a soffit, with much thicker panels on the vertical face. The coping and lintels are secured with stainless j-hooks and pins through holes in the webs of the hollow backed units (Figure 9.1.5). The buttress rising at the edge of the "dormer" form is made with large, hand pressed units that have unglazed recesses formed in the top and bottom for setting mortar, while the glazed faces terminate with a precise rim to be pointed later. The units are tied back to the concrete structure using the internal webs (Figure 9.1.6). The face of the building is accented with vertical pilasters. The molded edges of the pilaster were inspired by traditional shapes in the Boston Valley catalog. They were re-purposed in a vertical orientation and ultimately extruded with a new die. The complex pieces needed to transition the pilaster back to the building top were all hand pressed (Figure 9.1.7).

The façade techniques used at The Fitzroy invite comparison to those of the grand age of terra cotta in the 1920s. The sculptural potential of the material is the same. The methods of forming, finishing and attachment are still available. Approaches to modeling the original forms have changed with the use of digital tools. The overall management of production and inventory has been revolutionized.

A comparison can be drawn to Architect Timothy Pfleuger, who was a master of terra cotta use in Art Deco San Francisco.[5] His drawings and papers are preserved in the Bancroft Library at the University of California, Berkeley. He too made detailed pencil drawings of full building elevations that established the vertical emphasis, the step backs at the top, and the ornamental programs. His iconic Pacific Telephone Building (1925) was 26 stories tall, and therefore drawn at a small scale, but intentions for the detail were clear (Figure 9.1.8). The zones of intense ornament at the base and multiple tops were drafted again on large sheets of paper, 3 feet wide (91 cm) and 3 to 5 feet tall (91–153 cm). The details were developed through

[4] Boston Valley Terra Cotta, "The Fitzroy."
[5] Poletti.

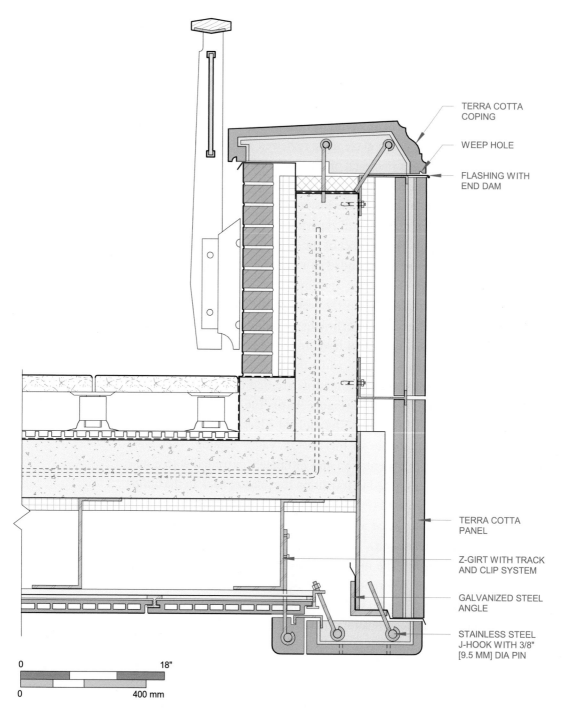

TERRA COTTA
COPING

WEEP HOLE

FLASHING WITH
END DAM

TERRA COTTA
PANEL

Z-GIRT WITH TRACK
AND CLIP SYSTEM

GALVANIZED STEEL
ANGLE

STAINLESS STEEL
J-HOOK WITH 3/8"
[9.5 MM] DIA PIN

0   18"

0   400 mm

Figure 9.1.5 Parapet detail at the 10th floor terrace. Extruded terra cotta panels, with custom, hand pressed coping. Fasteners include shelf angle, clips and stainless steel j-hooks with pins through the webs of the ceramic. From shop drawings by Boston Valley Terra Cotta.

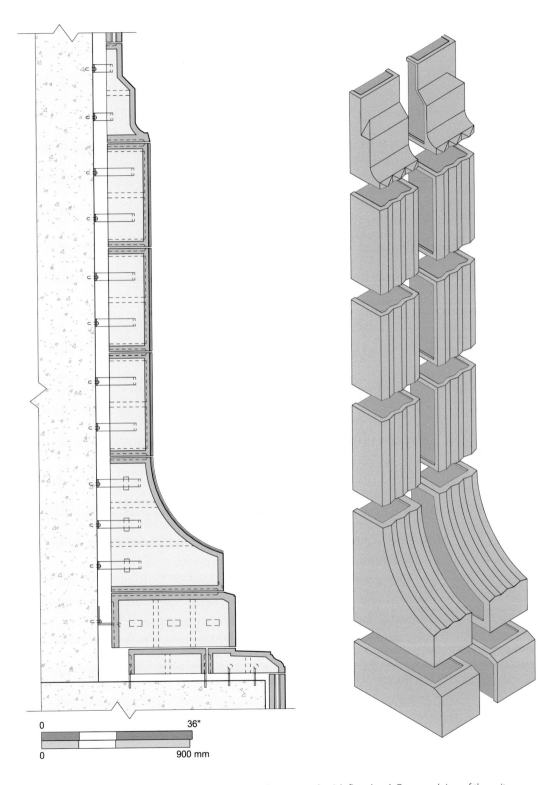

0 ——————— 36"

0 ——————— 900 mm

Figure 9.1.6 Hand pressed terra cotta blocks comprising a buttress at the 8th floor level. Faces and rims of the units are glazed. Recesses allow for setting mortar. Cross webs create hollow compartments at the back. Secured to the concrete structure with clips. From shop drawings by Boston Valley Terra Cotta.

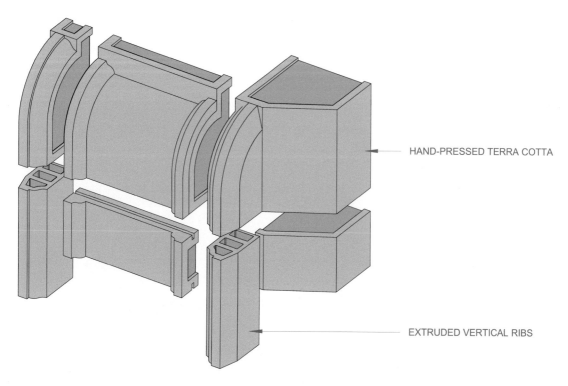

HAND-PRESSED TERRA COTTA

EXTRUDED VERTICAL RIBS

Figure 9.1.7 Termination of a projecting pilaster on the central façade at the roof level. Vertical ribs of extruded terra cotta from a custom die. Infill and transition elements hand pressed. From shop drawings by Boston Valley Terra Cotta.

multiple plan cuts, sections and elevations, integrated on a single drawing at ¾" scale (1:16). Draftsmen in the office produced full size sections (1:1) through the structure and cladding, showing the hollow cores of the terra cotta as well as the shelf angles, straps and j-hooks used to connect the blocks to the building. Today, the former would be worked out in 3D digital media, and the later would be described in shop drawings by the façade subcontractor and the terra cotta producer.

Pfleuger's drawings were sent to Gladding McBean, where the drafting department made meticulous scaled drawings of each piece and key drawings that located them on the façade. After approval by the architect, full size working drawings were prepared, enlarged to account for shrinkage of the clay after firing.[6] Unique details were developed in clay by sculptors in the modeling room at the factory. The completed models were placed on massive, tilt-up easels and photographed with a dry plate camera. Contact prints were made on matte paper so that the architect could mark them up with comments such as "fatten this" or "cut down" that. Copies of these photographs are among the papers of Gladding McBean in a special collection of the California State Library. They are in files consisting largely of laboriously typed correspondence documenting each piece with endless lists of inventory numbers packed into rail cars for shipment. Interspersed

[6] Kurutz, p. 92.

Figure 9.1.8 Large scale ornament on the upper floors of the Pacific Telephone Building, San Francisco, California (1925). Terra Cotta by Gladding McBean Company. Timothy Pfleuger, architect. Photo: Donald Corner.

are pleas to company staff on site that they urge the architect to come in person to the plant to confirm designs that could not be approved through photographs.

Custom terra cotta production at this scale essentially ended with the Great Depression and World War II. After the war, architects turned their interest to aluminum and methods of mass production that had been developed for the war effort. The epitaph for classical terra cotta was written in terms of all the repetitive, labor intensive steps required with costs that could not be sustained in the full employment years that followed. Many of these steps have been simplified or consolidated in the present, digital age. Although custom shapes are still smoothed by hand, automated conveyance, digital control of cutting or drilling, and robots able to apply glaze have reduced the labor component. The extraordinary potential of the material remains available to architects who are willing to look beyond preconceptions of style and fairly evaluate this new context.

Potsdamer Platz, Berlin, Germany (1994)
Renzo Piano Building Workshop

In 1992, the Renzo Piano Building Workshop (RPBW) and Christoph Kohlbecker won an invited urban design competition for a tract of land stretching from Potsdamer Platz, in what had been East Berlin, to the Kultureforum, in what was West Berlin. Historically, Potsdamer Platz had been a lively urban district, but it was cleared away for security margins along the transit of the Berlin Wall. The principal client for the redevelopment was Daimler-Benz Interservices, AG (Debis). RPBW designed many of the buildings in the scheme, starting with the Debis headquarters.[1]

The project became a benchmark for the contemporary revival of terra cotta. While it had been preceded by other RPBW projects and projects by Thomas Herzog, the scale and notoriety of Potsdamer Platz brought the material to the attention of architects around the world. Daimler required great precision in the ceramic elements, making the project a catalyst for a series of new production capabilities that reshaped the industry.[2] What follows is an account of the design process, written by Maurits van der Staay, a member of the team at RPBW and project architect for the façades.

• • • •

Renzo Piano Building Workshop was responsible for ten of the nineteen individual buildings projected in the winning masterplan. The other nine buildings were commissioned to the other participants in the competition. One premise for the design of façades, suggested in the masterplan, consisted in the use of mineral, ceramic or brick-like materials for the cladding on all of the buildings that were envisioned. This was intended to guarantee a level of coherence among these buildings.

This premise resulted in a variety of interpretations among the buildings, ranging from the dark red/brown of peat fired bricks on the Kohlhof tower, the pink-brownish ceramic tiling on the Arata Isozaki building, the reddish and yellowish sandstone on Rafael Moneo's buildings and to the

[1] Buchanan, 2003, p. 169.
[2] Lehmann, 2019.

DOI: 10.4324/9780429057915-12

reddish-yellow beige palette of the terra cotta extrusions on the building by Richard Rogers, Lauber & Wöhr and Renzo Piano Building Workshop.

When I joined the workshop in May 1994, and started working on the Potsdamer Platz projects, the use of ceramic extrusions was directly inspired by projects like the Banca Popolare at Lodi, and the Cité Internationale in Lyon, as they were recently finished or under construction.

At Lodi, the typical profile of the terra cotta extrusion corresponded to the layers of a traditional brick wall. The cladding on the buildings already featured baguettes with alternating voids, creating semi-transparent screens in front of windows. However, the extruded, ceramic elements were limited in length and required elaborate substructures to support them invisibly.

At Lyon, although longer wall panels were achieved, the "brise soleil" were composed of a series of shorter, baguette-like terra cotta elements (ca. 45cm) (1.48 ft), arranged as beads on a tube to attain the desired lengths

Façade Concept

The façade concept for the Debis-C1 building (the first to be erected) is that it be seen as part of a continuum stretching from the B1 building at the Potsdamer Platz on the north end, via the B2, B3, B5 and B7 buildings along the newly created Neue Potsdamerstrasse, to the C1 building on the south end at the Landwehrkanal.

This sequence forms a 600 m (1,968 ft) stretch of urban façades and proposes large portions of semi-transparent façade, in addition to the opaque claddings, in order to obtain gradual transitions between opaque, semi-transparent, and the fully glazed and double skin, glass façades.

The unifying characteristic over the entire length of these façades is the ceramic material.

It constitutes the outer layer of three that comprise the building envelopes:

1 The inner layer of the façades represents the physical enclosure of the building. As in the masterplan brief, and the constraints required by the Berlin municipal authorities, an overall ratio of 50% between opaque and glazed parts on all buildings was to be adopted. Furthermore, all buildings avoid the use of air conditioning by using natural ventilation through operable windows for ventilation and cleaning. In a typical unit, the glazing consists of two tilt and turn windows over one another, the upper one of which is motorized, to be opened automatically during summer nights to cool down the interior of the building.

2   An intermediate layer provides room to allocate blinds in front of glazing, or ventilated rock wool insulation in front of opaque concrete walls.

3   The outer layer of the façades is conceived as a continuous fleece of ceramic, uninterrupted by visible fixings that are instead concealed entirely within it.

On part of the buildings, like the southwest and south façades of the high-rise tower of the Debis-C1 building, and on the north end of the B1 high rise on the Potsdamer Platz, an additional layer has been added, the so-called double façades, in the form of tilting glass louvres, allowing windows to be kept open to ventilate the interior spaces even in windy conditions (Figure 9.2.1).

Design of the Unit and the Pieces

We started with the design and detailing of a basic unit for the C1 building, the first to be erected, and the double façade unit in front of it on the southwest side of the building.

In comparison to previous projects using terra cotta (we may do the same thing, but we never do it in the same way), the Potsdamer Platz façades posed a few major new challenges: longer and thinner ceramic extrusions, mullion covers in ceramic, and curved extrusions or pressed corner pieces.

Up until then, longer length hollow or profiled extrusions were commonly used mainly to cover the rock wool insulation on concrete walls. Typically, such profiles would measure 20 to 25 cm high (7.87 in – 9.84 in) by 3 to 5 cm deep (1.18 – 1.97 in) and come in lengths from 45 to 90 cm (1.48 – 2.95 ft). Since the C1 building was based on 8.10 m (26.57 ft) column grid, divided into six equal parts on the façade, a typical façade unit measured 1.35 to 1.38 m (4.43 – 4.53 ft) in width. With a floor to floor height of 3.75m (12.3 ft), the typical module measures 1.35m x 3.75m = 5 sqm (53.8 sqft).

This called for significantly longer extrusions than we had previously applied, up to 1.38 m (4.53 ft). The baguettes, which measured 48 x 48 mm (1.89 in) with a wall thickness between 8 and 12 mm (0.315–0.472 in) proved the most challenging to produce, in terms of straightness and regularity.

Each extruded horizontal profile contained at least one square tube in aluminum that provided residual security in case of breakage and allowed fixing to the vertical aluminum mullions of the façade modules.

As a second challenge, in the design of this module, we had to take into account that the cut edge of an extrusion in terracotta or ceramic has a

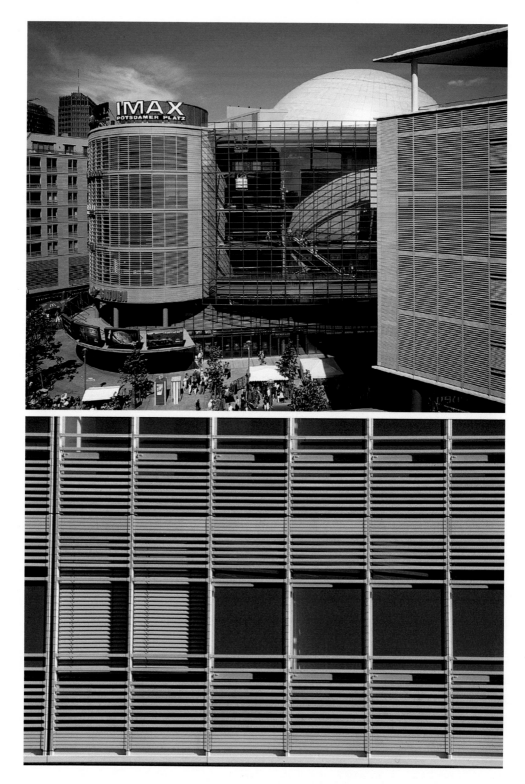

Figure 9.2.1 Potsdamer Platz redevelopment, Berlin, Germany (1994). Architectural design by Renzo Piano Building Workshop. Above: View from the south west toward the theater district in the central section. Photo: Michel Denancé. Below: Façade detail from the headquarters building for Daimler-Benz Interservices, AG (Debis). Photo: Gianni Berengo Gardin – courtesy of Fondazione per la Fotografia.

significantly different aspect than its skin, much like the crust of a bread differs from its interior. In most walls clad with ceramic or terracotta extrusions, these cut faces are either concealed by the metal frame in which they are integrated or left as an open joint of 3 to 4 mm (0.01–0.013 in).

In the final façade design, to prevent this cut edge from becoming visible, and to hide the actual mechanical fixing of pieces to the mullions from both the outside and the inside, we designed transition pieces in ceramic that would conceal both the aluminum profiles of the mullions and the connections of the extrusions to the mullions. This could not be done with straightforward extrusions but required form pieces that combined these two functions. For this we designed a series of "leibungsteile" (body parts), representing both window reveal and mullion cover, to be cast or form-pressed, rather than extruded (Figure 9.2.2).

At this stage, we had some kind of a unitized façade system in mind but imagined fixing the ceramic elements one by one on site. This was necessary to obtain what we now would call "reversibility": if for whatever reason an element anywhere in the façade should break or be damaged, it should be easy to replace it with a new one without having to tear down large parts of the façade (Figure 9.2.3).

We went to some length to study this approach and developed a "Faltblech" system to be able to rapidly fix and unfix each piece individually to the framework. (Faltblech refers to a "folded sheet" of stainless steel that would have provided a spring clip, allowing the baguettes to be snapped into place.) In the end, this was transformed by one of the brilliant engineers at Götz, Mr. Brosi, who conceived a machined piece of aluminum that connected the tubes to the mullion (Figure 9.2.4).

The third challenge regarded the rounded corner pieces. Since we wanted to create an uninterrupted veil of ceramic covering the buildings all around, these pieces should carry the same profiles around corners, and should be either bent extrusions or pressed pieces (Figure 9.2.5).

As far as the material itself is concerned, it had to be taken into account that the weather conditions in Berlin, compared to Italy or the South of France, can be quite extreme. Common terra cotta, as it is used in southern countries, tends to "breathe" more than the ceramics we eventually used, and is more susceptible to frost and the accumulation of dirt. This meant that the material itself had to comply with strenuous constraints regarding its strength, deformation and hygrometry.

On the basis of a preliminary façade design for the C1 façade we produced a mockup in pieces of painted wood to be installed on site in Berlin by December 1994, prior to approaching potential manufacturers of the material to verify their workability.

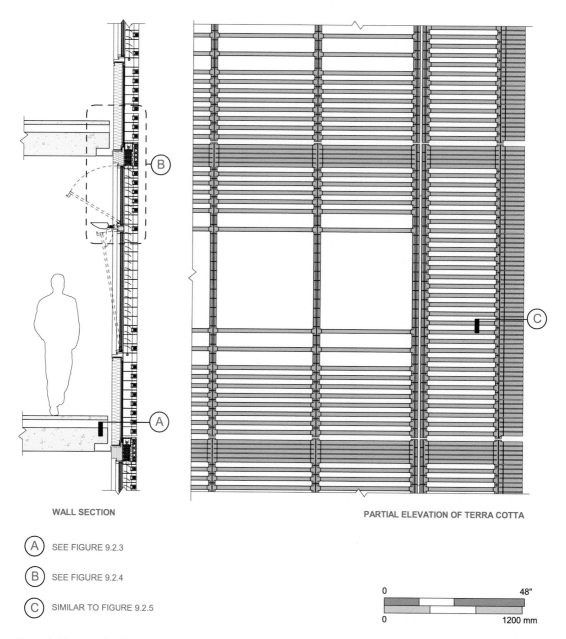

WALL SECTION

PARTIAL ELEVATION OF TERRA COTTA

(A) SEE FIGURE 9.2.3

(B) SEE FIGURE 9.2.4

(C) SIMILAR TO FIGURE 9.2.5

| 0 | | | | 48" |
| 0 | | | | 1200 mm |

Figure 9.2.2 Vertical wall section at the Debis Building, Berlin. A unitized, aluminum curtain wall system with exterior cladding and baguette screens of terra cotta by NBK Keramic, Emmerich, Germany. From drawings by Maurits van der Staay, project architect for the façades at RPBW.

Trial and Error: Test Pieces and Mockups

It was clear that the pieces we had imagined were not likely to be on anyone's shelf. They would have to be purposely conceived and made, but the sheer volume of façades justified the effort, approximately 70,000 sqm (753,474 sqft) on the RPBW buildings alone. We contacted several manufacturers to have their take on the feasibility of the extrusions and pieces we proposed, that for most of them were as new as they were for us.

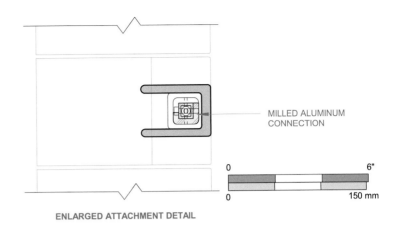

MILLED ALUMINUM
CONNECTION

0   6"

0   150 mm

ENLARGED ATTACHMENT DETAIL

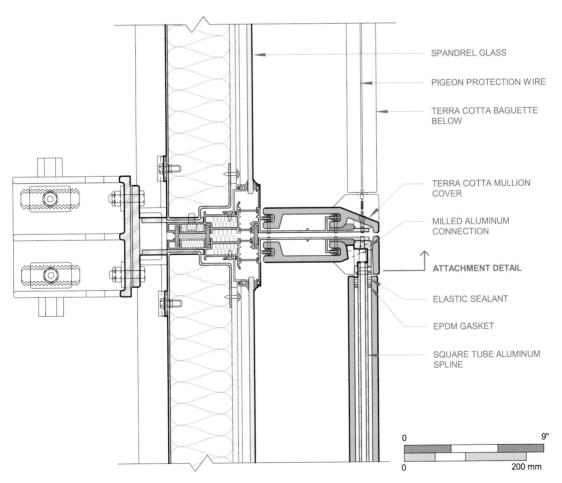

SPANDREL GLASS

PIGEON PROTECTION WIRE

TERRA COTTA BAGUETTE
BELOW

TERRA COTTA MULLION
COVER

MILLED ALUMINUM
CONNECTION

ATTACHMENT DETAIL

ELASTIC SEALANT

EPDM GASKET

SQUARE TUBE ALUMINUM
SPLINE

0   9"

0   200 mm

Figure 9.2.3 Plan detail of the typical mullion formed at the joint of adjacent curtain wall units. Custom terra cotta mullion cover forms a transition to extruded square baguettes. Aluminum details and fixings designed by Götz, GmbH, Wurzburg, Germany. From drawings by Götz and Maurits van der Staay.

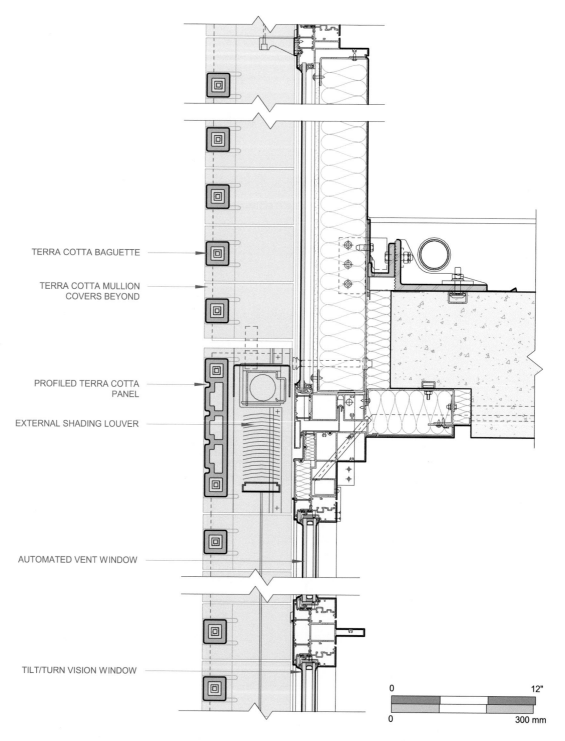

TERRA COTTA BAGUETTE

TERRA COTTA MULLION
COVERS BEYOND

PROFILED TERRA COTTA
PANEL

EXTERNAL SHADING LOUVER

AUTOMATED VENT WINDOW

TILT/TURN VISION WINDOW

0   12"

0   300 mm

Figure 9.2.4 Vertical section of curtain wall stack joint at the slab edge. Profiled terra cotta panels conceal external shading louvers. Baguette screens span the glazing, except in the vision zone. From drawings by Götz and Maurits van der Staay.

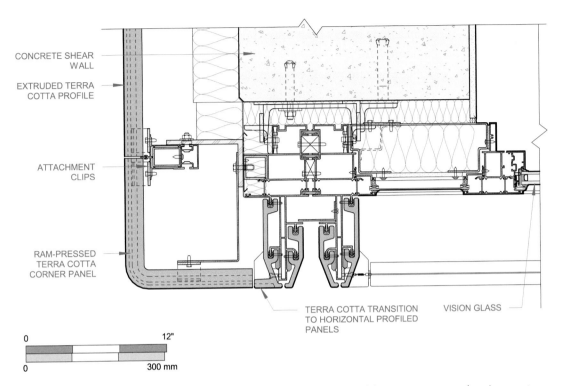

CONCRETE SHEAR WALL

EXTRUDED TERRA COTTA PROFILE

ATTACHMENT CLIPS

RAM-PRESSED TERRA COTTA CORNER PANEL

TERRA COTTA TRANSITION TO HORIZONTAL PROFILED PANELS

VISION GLASS

0   12"

0   300 mm

Figure 9.2.5 Plan detail of the corner transition from the curtain wall system, with baguette screens, to the rainscreen terra cotta cladding over solid, structural walls. Custom corner panel, ram pressed by NBK Keramik. From drawings by Götz and Maurits van der Staay.

Most of the firms had reservations on the size and complexity of the forms. One of the firms we consulted was Janinhoff, a mass manufacturer of brick and clinkers in Münster, Germany. It happened that they had just bought a small company on the Rhine near Krefeld, the Niederrheinische Baukeramik (NBK). This was a traditional firm, mostly specialized in crafty, decorative and bespoke extrusions and form pieces, that had started the production of longer extrusions for cladding purposes relatively recently. They provided us with some profiles and sample pieces of their standard products, that informed us on how to design the profiles for our project.

The type of ceramic they produce for extrusions is stronger and of a more sophisticated chemistry than plain terra cotta, which is closer to the straightforward firing of clay in a kiln, as has been done from ancient times. This chemistry allowed them to make a harder and more durable material that complies with the strenuous requirements of the Berlin climate and the building authorities. Their willingness and proactive motivation to experiment and innovate proved to be the basis of a fruitful collaboration, that lasted through the years, from the beginnings at Potsdamer Platz in 1994, up until the realization of the Saint Giles project in London, 2010.

For a second mockup, we established a catalog of individual extrusion profiles and pieces, corner pieces and mullion cover pieces, that were produced in the real material. We approached the design of the units and the individual pieces with great precision, necessary to control the sizes of joints and tolerances, and to set the tone of the quality we wanted to achieve. These pieces formed the basis on which we proceeded with the design of the rest of the façades on C1 and all the other buildings on Potsdamer Platz (Figure 9.2.6).

The Manufacturing Process

The making of a piece or an extrusion requires several consecutive steps:

1　We produced a very precise Autocad drawing of the desired element, in section and elevation in scale 1:1, including wall thicknesses and radii.

2　At NBK this drawing is redrawn to make the mouthpiece for the extrusion, or the counter-form for pressed pieces. Hereby, the architects' drawing is scaled up to take into account the shrinkage factor that occurs during the drying and firing. Depending on the composition of the grog, this is usually anywhere between 5 and 7%. The mouthpiece (die) consists of laser, or water cut, steel plate of 20–25 mm (0.79–0.98 in) thickness, through which the clay-mass is pressed.

3　The preparing of the grog: a mix of ingredients, similar in chemical complexity to Coade stone, or other ceramic-gres like materials, subject to various stages of fine tuning, whereby the proportion of the ingredients, such as different types of clay, chamotte, quartz, water and oxides is adjusted according to the result of multiple trial runs.

4　The extrusion of the pieces, whereby plastic clay is pressed through the mouthpiece, cut to rough length with a wire, and lands on a conveyor belt that transfers it on to a drying rack.

5　Drying. The extruded or pressed pieces are held in an environment that starts at room and finishes the drying process at about 70°C (160°F), simultaneously reducing the humidity of the air in the chamber. The time for this process is 48 to 72 hours depending on the size and shape.

6　Firing. Once dried, the extrusions and/or pressed pieces are transferred into a kiln, to be fired at approx. 1100°C (2000°F) for 24 to 72 hours.

7　Cooling down to room temperature.

8　Cutting to size, and eventually, drilling.

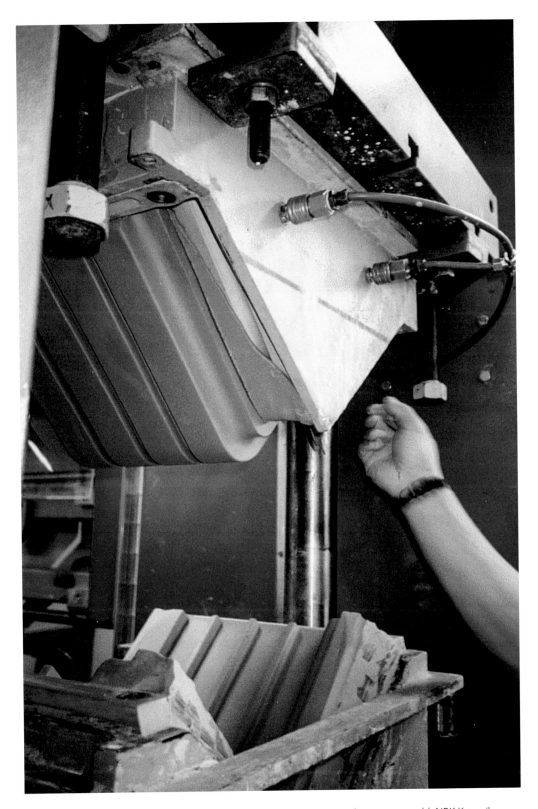

Figure 9.2.6 Forming the custom, ribbed corner panels by ram pressing with a two part mold. NBK Keramik, Emmerich, Germany. Photo: © Maurits van der Staay.

9 Glazing (as we did for the Saint Giles project). For each color, a liquid glaze slurry containing a mix of vitreous substances, oxides and pigments is prepared and brought on the already fired and cut pieces, for a second firing. These formulated glazes vary in viscosity and require individual firing temperatures.

10 Customizing the pieces to be integrated in the façade units, like integrating tubes and fixings.

As may be clear, this is an iterative process, and each step of fabrication requires adjustment of any of its parameters: composition, humidity, viscosity, temperature and timing. In the beginning, many pieces had to be discarded, due to unwanted deformations and cracks that appeared during drying and/or firing. This amounted to 25% in the beginning, to be reduced to 5% during the final production phase. Nevertheless, most of the discarded material is recycled and finds its way into the grogs used in future batches.

From this beginning we established a regular exchange with NBK, to design and produce test pieces, adapt them and create new pieces. During this process of trial and error, we had the opportunity to get familiar with the production in great detail, over almost a year producing prototypes and tests to achieve the desired lengths, wall thickness and further material properties, like straightness, color and surface quality.

The design of the façades on the C1 building led to a great number of derived typologies and pieces for the subsequent buildings designed by RPBW for Potsdamer Platz during the period that followed. These vary from fully glazed façades (wedges, curves, roofs, south staircases), to opaquely clad façades (back of house, chimney, corridors and side-walls). In between are all the elements that are partly clad with ceramic pieces, like the typical units for the C1 (Debis) offices, the adjacent residential units, the rotunda and the wedge.

Ultimately the pieces included flat faced extrusions and ribbed face extrusions. These were both curved and straight. The baguettes were also curved and straight. The transitional mullion cover pieces were used on five buildings. There were flat faced and ribbed corner pieces at a variety of angles: 90°, but also 35°, 43°, 53°, 80°, 100°, 127° and 145° (see Figure 9.2.5).

Assembly of Façade Units and On-site Installation

As we moved into the tender phase, initially only for the Debis C1 building, working with Götz GmbH from Würzburg in Germany, it became clear that in view of the size of the job we would build the façades in a modular, unitized system, whereby the elements would be prepared off site, to be mounted one by one on site, to constitute the entire envelope of the buildings.

This would enhance the speed of installation and reduce the number of operations to be performed on the site itself to a minimum (Figure 9.2.7).

Moreover, the use of CNC machining allows for great precision (to the fraction of a millimeter) and allows variation and customization of profiles within the façade units. This guarantees the accurate assembly of 62 to 82 precisely cut ceramic parts on frames in the regulated environment of a workshop. The units as a whole could be installed, in turn, with the same precision on site.

In this constellation, NBK produced the ceramic extrusions and pieces, to be shipped to Götz's workshop in Deggendorf, Bavaria, where they were assembled on the units that contained all the other elements, like windows (including their motorization), an insulated spandrel panel, sunshading slats, etc. These fully prepared floor to floor units, 3.75 x 1.35m, were then shipped to Berlin to be installed by crane. On the C1 building alone, this amounted to more than 5,000 individual units.

Figure 9.2.7 Components of the terra cotta cladding and baguette screens. Top left: Square aluminum tubes positioned in the center of the extruded baguettes using gaskets. Bottom left: Installing segments of mullion cover on the vertical stiles of the curtain wall unit. Right: Two views of the transition pieces joining the baguettes to the mullion covers. The milled aluminum seat receives the square tube in the center of the baguette. Photos: ©Maurits van der Staay.

Simultaneously, we continued the process of developing the façade elements for the other buildings with both NBK and Götz.

During this period, in view of the projects following the Debis C1 building, NBK invested in the modernization of its equipment. The old, coal powered kilns were gradually replaced with gas fired ovens of around a 100 m (328 ft) in length, in which the pieces traveled at slow speed (24 hours for 100m) during firing. Drying facilities, mixers and the laboratory all were brought up to a match with the huge volume required.

Each curtain wall unit produces 3.75 x 1.35 = 5 sqm (53.8 sqft) of envelope. The ceramic coverings range as follows:

Fully opaque units: 16 flat faced extrusions
Standard office units of C1: 62 individual pieces
50% shaded units: 87 individual pieces

The total, calculated across 70,000 sqm (753,474) of building façades, corresponds to roughly 546,000 pieces.

Maurits van der Staay, Paris, August 2020

Central Saint Giles, London, UK (2010)
Renzo Piano Building Workshop

Continuing below is an account of the design process at RPBW, written by Maurits van der Staay who was Associate in Charge for Central St. Giles.

• • • •

This 60,000 sqm (645,835 sqft), mixed use development sits on an island site of 7,000 sqm (75,347 sqft), in the heart of London's historical center. There was the desire to make a permeable and accessible place, and a crafted building, responding to the irregularity and diversity of its surroundings. In order to do this, we raised the ceiling height of the ground level to 6 meters (19.68 ft) and strived to make this level as transparent as possible, reducing the size of the cores and loading bay to the minimum necessary, allocating most of the kitchens and toilets to the basement, and applying full height, extra clear glazing all around. This created a piazza in the middle of the site, completely transparent, visually and physically permeable through five passageways. This was in response to the uninviting and somewhat gloomy building that occupied the site at the time (Figure 9.3.1).

In order to break up the bulk of this new building, the façades were divided into twenty-one slightly orientated "faces," in dialogue with the irregular orientations of the surrounding buildings. They alternated with deep incrustations, setbacks and winter gardens, all fully glazed. The surfaces of these faces were to "vibrate" in resonance with varied brick and stone façades of buildings around the site.

The individuality of the faces was further enhanced by giving them different colors, inspired by the often brightly colored surfaces that appear on shop fronts and façades in the typical mews and yards tucked away in the London urban fabric. With the use of glazes, the ceramic seemed to lend itself perfectly to both purposes, and we could build upon the experiences obtained at Potsdamer Platz (Figure 9.3.2).

DOI: 10.4324/9780429057915-13

Figure 9.3.1 View into the "piazza" of the Central St. Giles mixed use development, London, UK (2010). Building façades composed of terra cotta clad, unitized curtain walls. Designed by Renzo Piano Building Workshop with Maurits van der Staay, Associate in Charge. Photo: John Rowell.

Figure 9.3.2 The outward looking façades at Central St. Giles are broken down into defined sections, including slight changes in orientation and distinctive glaze colors. Terra cotta by NBK Keramik. Photo: © Maurits van der Staay.

In Berlin, as at Lodi and most of the previous projects, the ceramic pieces were arranged flush against an outer boundary. The only protruding elements were attic and portico cornices, and the façade surfaces could thus be considered as a "counter relief," moving toward the inside of the volumes. At the Saint Giles façades, instead, protruding elements extend toward the outside, creating a raised relief, similar to the surfaces of the brick and stone walls on the buildings in front of them. In order to create this depth, we adopted a kind of weaving technique, which we tried out in a great number of models before preparing a series of life-size mockups (Figure 9.3.3). To build these, we adopted a similar process of exchange with NBK as we had grown familiar with them during the Berlin years.

Unlike the Potsdamer Platz façades, where a great number of pieces were pressed and customized by hand before firing, at Saint Giles the ceramic parts on the units are almost entirely made with extrusions, except for the plug pieces that close off the exposed hollow ends of baguettes. For the Saint Giles project, we developed the elements with Schneider

Figure 9.3.3 Relief in the façades generated by building outward through multiple layers of terra cotta. Left: The weaving together of horizontal and vertical profiles and baguettes. Right: A connection detail showing an outer layer of baguettes, back-fastened through an inner layer. Photo: © Maurits van der Staay.

Fassaden from Stimpfach, Germany. The off-site mass preparation of the 3,306 façade elements that constitute the building envelope took place in a factory in Wroclaw, Poland, before being transported to London. Unlike the Berlin projects, where the units were mainly installed with cranes and scaffoldings, installing was done with so-called "manipulators," allowing the façades to be mounted from the inside of the building, with a minimum impact on other trades working on site simultaneously.

Maurits van der Staay, Paris, August 2020

• • • •

Central Saint Giles remains a landmark project for the fine grained, multi-layered application of terra cotta. Maurits van der Staay was Associate in Charge at Renzo Piano Building Workshop. The weaving of extruded sections described above was developed through sketches, drawings, models and prototype pieces (Figure 9.3.4). There were mockups built at the Workshop in Vesima, Italy and on site in London. The early sketches explored connection strategies that would allow a layer of baguettes to be connected though the pieces passing behind or in front of them (Figure 9.3.5).

Figure 9.3.4 A finely crafted model of the St. Giles façade system demonstrating the relief in the terra cotta layers. Part of a RPBW traveling exhibition on display in Padova, Italy, May 2014. Photo: Donald Corner.

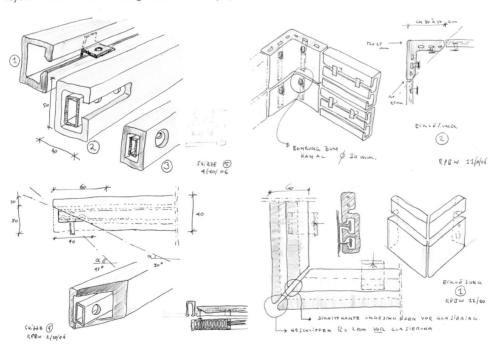

Figure 9.3.5 Design development sketches by Maurits van der Staay, exploring alternative configurations for the St. Giles components. Left: Connection strategies at the ends of the baguettes. Right: Different approaches to the problem of turning the outside corner. Photos: © Maurits van der Staay.

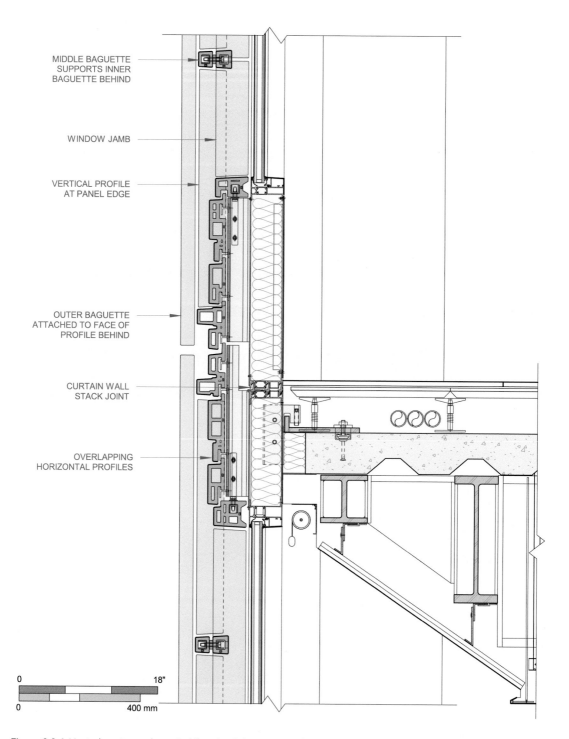

MIDDLE BAGUETTE
SUPPORTS INNER
BAGUETTE BEHIND

WINDOW JAMB

VERTICAL PROFILE
AT PANEL EDGE

OUTER BAGUETTE
ATTACHED TO FACE OF
PROFILE BEHIND

CURTAIN WALL
STACK JOINT

OVERLAPPING
HORIZONTAL PROFILES

0   18"

0   400 mm

Figure 9.3.6 Vertical section at the typical floor level showing a stack joint between two units of prefabricated curtain wall. The layers of terra cotta were applied off site. From drawings by Maurits van der Staay.

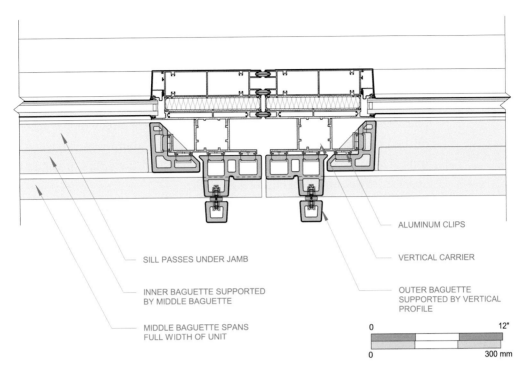

ALUMINUM CLIPS

SILL PASSES UNDER JAMB

VERTICAL CARRIER

INNER BAGUETTE SUPPORTED
BY MIDDLE BAGUETTE

OUTER BAGUETTE
SUPPORTED BY VERTICAL
PROFILE

MIDDLE BAGUETTE SPANS
FULL WIDTH OF UNIT

0   12"

0   300 mm

Figure 9.3.7 Plan detail at the meeting of adjacent curtain wall units. Vertical aluminum carriers are notched to allow the terra cotta sill to pass under the jamb. Profiles are overlapped to prevent a direct view through the terra cotta to the backup structure. From drawings by Maurits van der Staay.

Figure 9.3.8 Two elevation views at Central St. Giles demonstrating the design intent. Left: The texture of light and shadow created by layers of baguettes and profiles. Photo: © Maurits van der Staay. Right: Coordination of glaze colors with the natural and architectural features in the context of the project. Photo: ©RPBW, Francesca Bianchi.

The spandrel zone of the wall units picks up the theme of weaving with projecting profiles that match the dimension of the baguettes. The solid background is banded with sections designed to overlap so there are no open lines of site through the terra cotta. A reversible piece at the head and the sill covers the aluminum frame holding the glass (Figure 9.3.6). The plan detail shows these sill pieces sliding under the jambs, which is accomplished by cutting away a portion of the vertical aluminum carrier. Also indicated are color coordinated spacers that hold the layers of baguettes apart from one another (Figure 9.3.7).

Glazing at the office spaces is fixed. For the housing, there is a glass guardrail in front of operable units, as can be seen on the right edge of Figure 9.3.2. The window jambs are extended to reduce the glazing area and the baguette weave is simplified.

All of this attention to layering pays dividends in the play of light and shadow across the façades. The colors on the outward faces of the complex were tuned to avoid a cold, or "chemical" quality, and to resonate with the context (Figure 9.3.8).[1]

[1] Van der Staay.

125 Deansgate, Manchester, UK (2021)
Glenn Howells Architects

Glenn Howells Architects have designed a contemporary, "Grade A" office building at 125 Deansgate in the Manchester city centre (UK). Deansgate is an historical, commercial street connecting the city north to south. The building forms a prominent corner at Brazenose Street; a pedestrianized way that connects west to east from the financial quarter at Spinningfields to the Civic Quarter, that contains a collection of both historic and modern commercial, cultural and civic structures. Heritage buildings in the area are made of "Manchester Red" bricks with terra cotta details. Other, Victorian era buildings feature glazed terra cotta backed up with masonry. Directly across Deansgate is the John Rylands Library, a "Grade 1" listed building of solid construction in red sandstone, with fluted tracery in the principal window details. The choice of unglazed, red terra cotta as cladding for the new building was a strong response to the context, and through careful detailing it fulfills a dialogue between the historic and modern structures in the area (Figure 9.4.1).

The new building provides 115,000 square feet (10,684 sqm) of office space and 12,000 square feet (1114.8 sqm) of retail space at street level. Consistent with the context, the façades have a tripartite organization with a two-story base, a two-story top and eight stories in between. Louis Sullivan's landmark Guaranty Building in Buffalo, New York was also cited as a reference by the design team. The office floors are laid out on a 1.5 meter (4.92 ft) planning grid for flexible use and those dimensions carry on to the development of the cladding. The steel frame structure spans from a stout core directly to a lattice of slender columns on a 3 meter (9.84 ft) grid at the perimeter.

Key concepts for the façade include a strong expression of the vertical with a sense of solidity, limited by the need to maximize daylight entry to the office floors. Responding to the richness of the adjacent masonry

DOI: 10.4324/9780429057915-14

Figure 9.4.1 View of 125 Deansgate, Manchester, UK (2020), at the corner of Brazenose Street. Historic John Rylands Library immediately to the left. Glenn Howells Architects. Terra cotta fixed to the unitized curtain wall articulates paired floors at the bottom, top and middle of the building. The south façade (right) has a closer spacing of major verticals. Photo: Greg Holmes on behalf of Glenn Howells Architects.

buildings, an early goal was to achieve a "metal free façade" with the terra cotta juxtaposed directly to the glass.[1]

The preliminary design called for composite terra cotta profiles backed up with reinforced concrete. The joints were to be grouted to express traditional solidity. Depth in the façade was desirable to create a play of shadows across the surface and to cut off oblique sun angles. The façade contractor favored a unitized curtain wall approach for extensive pre-assembly and rapid installation. The curtain wall units themselves provide the thermal, air and water barriers for the building with open jointed terra cotta elements mounted to the exterior.

The Deansgate façade, facing west, is composed in plan with 625 mm (24 in) terra cotta piers in front of the column positions and 160 mm (6 in) terra cotta fins in between (Figure 9.4.2). After a series of design trials, the profile selected for the piers has a scalloped shape derived from the traditional stone mullions of the John Rylands Library. The intermediate fins have a similar shape at a reduced scale. The piers are composed of three extruded elements that are back-fastened to aluminum trays with

[1] Glenn Howells Architects.

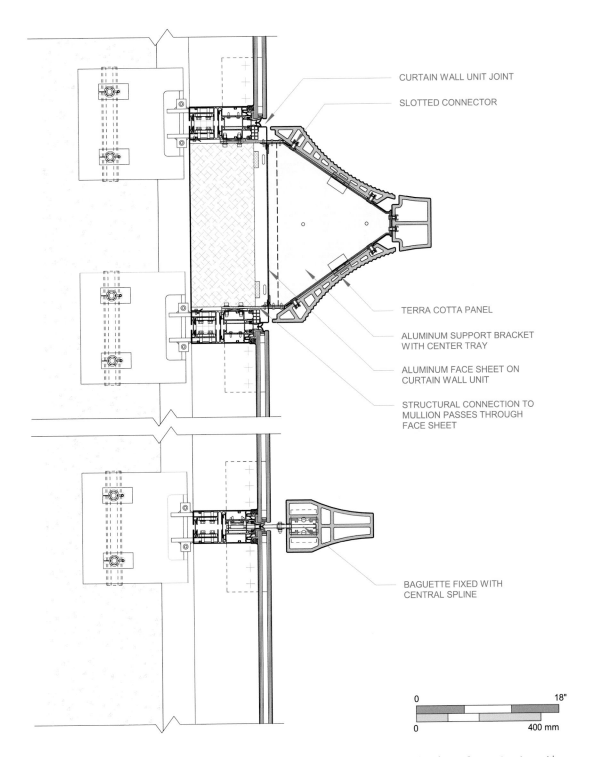

CURTAIN WALL UNIT JOINT

SLOTTED CONNECTOR

TERRA COTTA PANEL

ALUMINUM SUPPORT BRACKET
WITH CENTER TRAY

ALUMINUM FACE SHEET ON
CURTAIN WALL UNIT

STRUCTURAL CONNECTION TO
MULLION PASSES THROUGH
FACE SHEET

BAGUETTE FIXED WITH
CENTRAL SPLINE

0   18"

0   400 mm

Figure 9.4.2 Plan detail of vertical terra cotta elements in the central portion of the façade. Above: Composite piers, with terra cotta elements attached to brackets that span the full width of a curtain wall unit. Below: Extruded fin attached to one edge of a glazed curtain wall unit. Terra cotta by NBK Keramik. From drawings by Glenn Howells Architects and Rinaldi Structal, Colmar, France.

folded edges. These trays are attached through the sheet metal facing of an opaque aluminum curtain wall unit that exactly matches the pier in width. The sheet metal face and the gasketed joints of the unit complete the physical enclosure of the building. These curtain wall units span the 4.0 meter (13.12 ft) floor to floor height, and each one is pre-fitted with three vertical lifts of the terra cotta profiles before placement. The adjacent unit is fully glazed with the smaller terra cotta fin attached to one edge. This is followed by another glazed unit and then another pier. The Brazenose façade, oriented to the south, has a slightly reduced glazing ratio to control solar gains. This is accomplished by repeating the composite pier on 1.5 meter (4.92 ft) spacing (Figure 9.4.3).

The vertical proportions of the building are controlled by introducing a horizontal sill and spandrel panel at the slab lines, to contrast with the powerful, vertical piers (Figure 9.4.4). After study of the overall composition, it was decided to use the spandrel detail at alternate floors to create a double story proportion that harmonizes with the double top and double

Figure 9.4.3 Left: Assembly of the story height curtain wall units with terra cotta spandrels, piers and fins previously attached. Photo: Rob Parish. Right: A composite pier mounted to the face of an entire curtain wall unit, leaving no metal visible to the exterior. Terra cotta by NBK Keramik. Photo: Glenn Howells Architects.

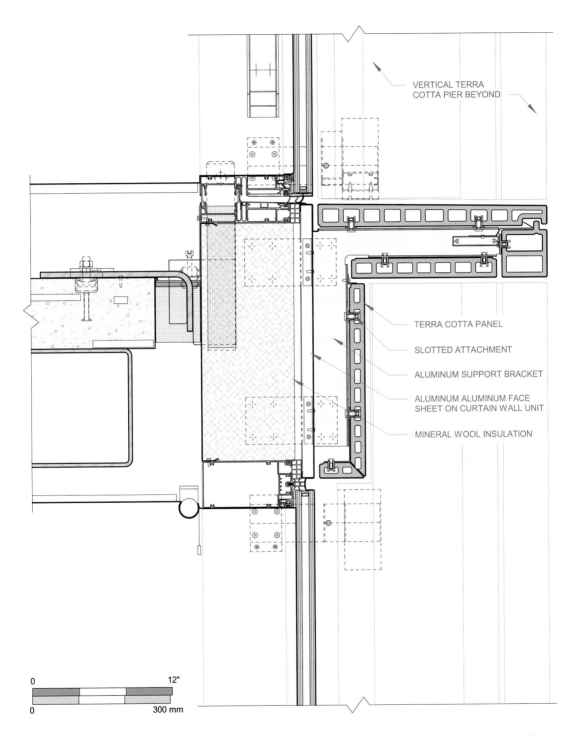

VERTICAL TERRA
COTTA PIER BEYOND

TERRA COTTA PANEL

SLOTTED ATTACHMENT

ALUMINUM SUPPORT BRACKET

ALUMINUM ALUMINUM FACE
SHEET ON CURTAIN WALL UNIT

MINERAL WOOL INSULATION

0   12"

0   300 mm

Figure 9.4.4 Vertical section of the spandrel panels applied at two story intervals to provide scale and proportion in the central zone of the façade. Terra cotta supported by aluminum brackets that pass through a metal weather skin on the curtain wall unit. From drawings by Glenn Howells Architects and Rinaldi Structal, Colmar, France.

base. There is a nearly seamless stack at the intermediate floors. The vertical terra cotta and the glass flow continuously upward with a reflective metal panel attached to the inside edge of the aluminum frames to conceal the floor assembly.

The piers and fins create a 514 mm (20.24 in) deep screen overlaying the glass plane of all the exposed façades. The respective horizontal bay layouts are terminated with a major pier on each side of the corner (Figure 9.4.5). This

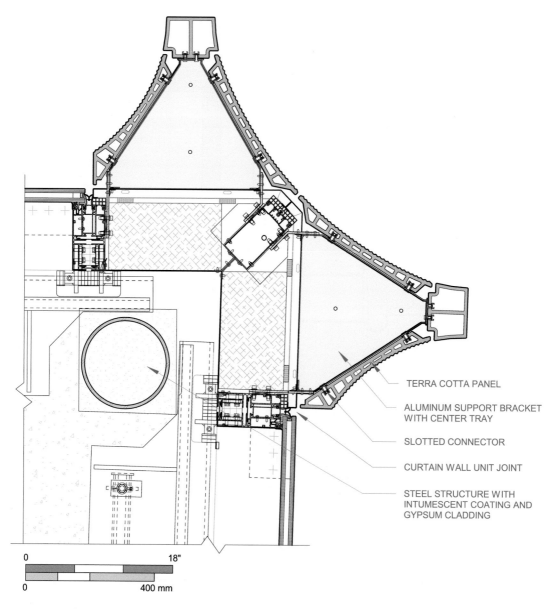

TERRA COTTA PANEL

ALUMINUM SUPPORT BRACKET WITH CENTER TRAY

SLOTTED CONNECTOR

CURTAIN WALL UNIT JOINT

STEEL STRUCTURE WITH INTUMESCENT COATING AND GYPSUM CLADDING

0   18"

0   400 mm

Figure 9.4.5 Plan detail at the building corner. Two piers are attached to a corner curtain wall module, each one terminating the layout of its respective façade. From drawings by Glenn Howells Architects and Rinaldi Structal, Colmar, France.

logical and elegant detail expresses the wholeness of each façade overlay and gives the building characteristic strength at the outside corners.

The important urban intersection on the southwest corner is further reinforced at the base with a double height colonnade opening toward Spinningfields Square. The intersection between the central shaft and the base is marked with a larger assembly of horizontal shelf, cove and beam wrap appropriate to the column spacing that is three times that of the wall above. The assemblage is made up from extruded shapes that fall within the dimensional range available from the terra cotta provider. Again, they are back-fastened to folded sheet metal pans that are, in turn, attached to the frame of the curtain wall units (Figure 9.4.6). The soffit of the colonnade is composed of terra cotta panels attached to the underside of the structure. Blind fasteners fit into keyhole shaped slots, allowing these larger pieces to by mounted on site after the curtain wall and the support armature have been placed. Further north along Deansgate, shop front glazing steps forward to the façade plane as an expressed infill condition, allowing the same terra cotta details to continue across the full width of the building.

The top of the building is articulated with a double height terrace, recessed behind slender columns, 8 meters (26.25 ft) tall. With a width of only 300 mm (11.8 in) and depth of 735 mm (28.93 in), these columns are a four-sided wrap of terra cotta profiles around a tube steel armature that provides lateral support. There were each installed in one piece, pinned to the cantilevered edge of the roof structure above and the floor structure below.[2]

The architects collaborated with staff at NBK Keramik GmbH, in Germany, to develop the terra cotta profiles. The curtain wall units were fabricated by Rinaldi Structal of France. The closely spaced steel frame with lightweight, castellated floor beams produced a structure with very little movement. As the terra cotta system was refined, it was possible to reduce the open joints from a 25 mm (0.98 in) spacing to only 12mm (0.47 in). These tighter and more consistent joints helped realize the sense of solidity for a building in this context. The greatest challenge came in resolving the crossing joints between the horizontal and vertical elements of the central façade.

In summary, 125 Deansgate is a direct and highly refined essay on three formative materials: steel, glass and terra cotta. There is a generous share of extruded aluminum to hold the façade together, but it is not exposed to the exterior. The vertical expression of glass and terra cotta resonates with its antecedents, the revolutionary skyscrapers of Chicago, first introduced in Chapter 2. Thomas Leslie's book (*Chicago Skyscrapers, 1871–1934*) traces the development of the curtain wall,

2 Desborough and Shambi.

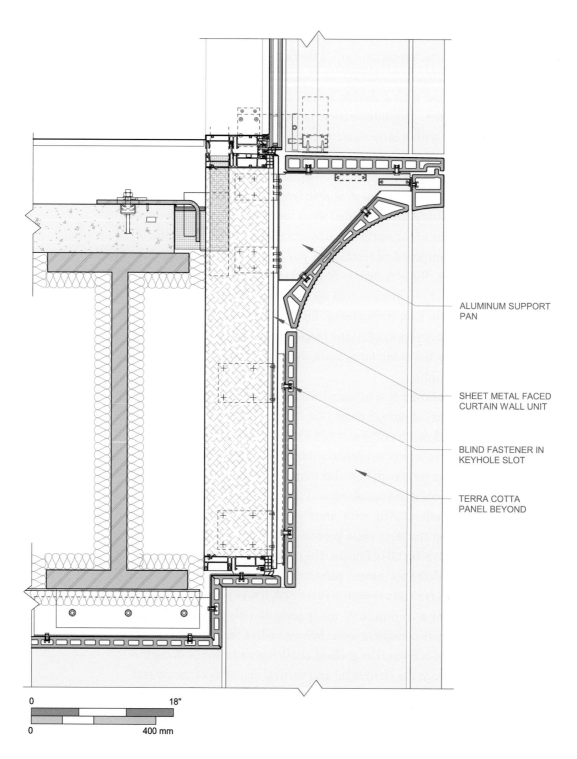

ALUMINUM SUPPORT
PAN

SHEET METAL FACED
CURTAIN WALL UNIT

BLIND FASTENER IN
KEYHOLE SLOT

TERRA COTTA
PANEL BEYOND

0   18"

0   400 mm

Figure 9.4.6 Vertical section of the spandrel across the top of the commercial colonnade at street level. Terra cotta units field applied at the end of the construction sequence. From drawings by Glenn Howells Architects and Rinaldi Structal, Colmar, France.

leading up to a pair of landmark buildings by D.H. Burnham and Company: the famous Reliance Building of 1891–5 and the less known Fisher Building of 1896. The Fisher Building makes a stronger comparison, with street façades on three sides of a 75 x 100 foot (22.86 x 30.48 m) site and clear spans from a central circulation corridor to columns at the perimeter.

The Fisher Building was introduced as a "building without walls" in a contemporaneous technical review.[3] As with Deansgate, the designers sought to harvest the maximum amount of daylight for the interior spaces. The lightweight steel girder system, with moment joints, freed the exterior for a broad expanse of glass.[4] At the time of construction, there was an excess supply of plate glass available from massive new Indiana factories, providing an economic incentive to push the glazing area as far as possible. Deansgate does the same thing for only slightly different reasons, capitalizing on the fabrication and placement efficiencies of a unitized curtain wall system to cover as much of the façade as possible.

Curtain wall construction in the 1890s startled the populace because it no longer required construction from the ground up. The highly regularized claddings in the middle of the building could be manufactured and applied first, leaving the base open for the delivery of materials at an urban site. The specialized components at the base were installed at the end when they were no longer at risk of damage due to the construction above.[5] Construction photos of the Fisher Building show what was then a novel sequence, but now common practice.[6] Photos of Deansgate are very similar with the curtain wall proceeding upward as the base awaits the field assembly of larger units of infill glazing and terra cotta structural wraps, only when it is safe to install them.

The strongest parallel between the two buildings is the timeless quality of the terra cotta. The material was favored in the 1890s because it could be extruded with great precision and drawn very closely around the columns and spandrels of the steel frame.[7] This tight wrapping saved space on the floor plate, opened the façade to light, and powerfully expressed the slender structural grid on the exterior of the Fisher Building (Figure 9.4.7).

At Deansgate, the terra cotta is no longer required to fireproof the steel, as that is accomplished with intumescent coatings and gypsum cladding. However, the elegance of the building derives from that same ability of terra cotta to cover only those slender sections that need covering. The repeating, extruded profiles produce strong vertical lines, complemented by understated horizontals. Light and shadow play across the subtle shapes of the piers and fins, shapes that derive from a logical and efficient approach to working with clay (Figure 9.4.8).

3 "Technical Review, The Fisher Building."
4 Leslie, pp. 91–3.
5 Leslie, p. 79.
6 Leslie, p. 96.
7 Leslie, p. 86.

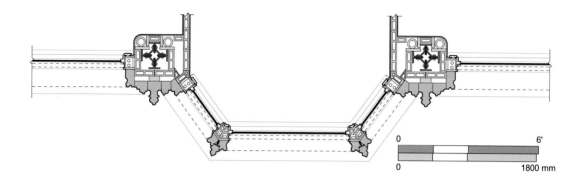

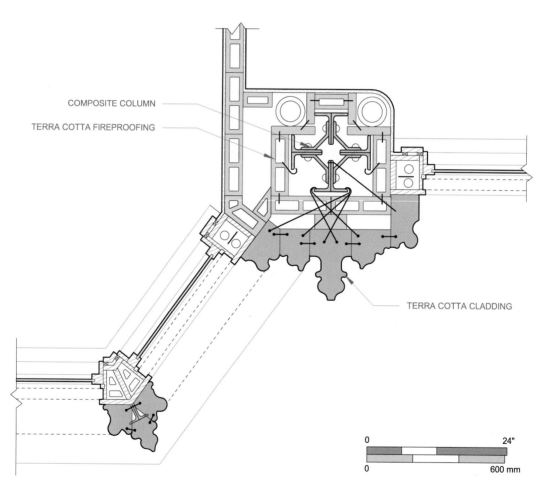

COMPOSITE COLUMN

TERRA COTTA FIREPROOFING

TERRA COTTA CLADDING

0   24"
0   600 mm

Figure 9.4.7 Fisher Building, Chicago, Illinois (1896). Plan details of the characteristic oriel windows. The lightweight, composite structural frame is fire protected with terra cotta inside and out. Source: "Technical Review, The Fisher Building".

Figure 9.4.8 View toward the corner at 125 Deansgate, with the sandstone of John Rylands Library to the left. Unglazed, red terra cotta harmonizes with the context. Photo: Greg Holmes on behalf of Glenn Howells Architects.

Tykeson Hall, Eugene, Oregon (2019)
Office 52 Architecture + Rowell Brokaw Architects

Tykeson Hall opened in September 2019 at the University of Oregon, in Eugene. Located on an infill site at the heart of the historic campus, it provides an administrative center for the College of Arts and Sciences as well as innovative program space for student success. The building was designed by Office 52 Architecture, Portland and Rowell Brokaw Architects, Eugene.

The building is expressed as three interlocking volumes: state of the art classrooms; the student services zone; and common areas. From the beginning, it was intended that these spatial/functional distinctions be carried through to the materials. Brick for the classrooms and glass for the commons derived from the surrounding buildings and connection to a new campus open space, respectively. The third volume is a multicolored terra cotta rainscreen, because the client group requested a quality material that addressed the sensitivity of the site and the uniqueness of the program. Adjacent buildings integrate brick and terra cotta through historic styles. At Tykeson, they are independent expressions of high performance cladding, one traditional and the other contemporary (Figure 9.5.1).

The project budget was adequate, but terra cotta represented a significant upgrade over the metal and cementitious panel products originally in the scope. The limited number of qualified subcontractors available and the relatively small size of the project further constrained the terra cotta system. The result is a high quality outcome that is very nearly an off-the-shelf application of the material, unlike the examples in the previous case studies.

The façades are composed in modules of one, four and eight feet (30.5, 122, 244 cm). For maximum economy, there is a single extrusion, producing a nominal six inch (15 cm) tile inclusive of 3/8 inch (9.5 mm) open gaps. The mitered corners are milled from the same section (Figure 9.5.2). There are five custom colors and a carefully limited set of tile lengths.

DOI: 10.4324/9780429057915-15

Figure 9.5.1 Tykeson Hall, at the University of Oregon (2019). Three interlocking volumes express the building functions: common areas (glass) connect to the campus, classrooms (brick) relate to the adjacent buildings, and student services (terra cotta) occupy the central position. Office 52 Architecture and Rowell Brokaw Architects. Photo: Frank Visconti for Rowell Brokaw Architects.

The brick cavity wall and the terra cotta rainscreen are closed at the window penetrations with tapered aluminum mullion caps, supplied with the glazing. Both systems are detailed as "perfect walls" with attention to control layers and thermal breaks (Figure 9.5.3).

The terra cotta was supplied from Europe by Shildan. The original submittal called for overlapping end cuts on the vertical tiles. Since the tiles are single glazed, this would have exposed the body color at the back of the joints. Instead, the tiles were square cut at intermediate joints and the aluminum clips were painted black on site. A continuous black scrim conceals the bulk of the unpainted support structure (Figure 9.5.4).

The size and location of the project relative to established markets presented challenges despite the discipline that was imposed on the system. Color approval was based on small square chips, not full tiles (Figure 9.5.5). There was no full-scale mockup as all of the terra cotta was shipped at the same time, arriving twenty-nine weeks after confirmation of the

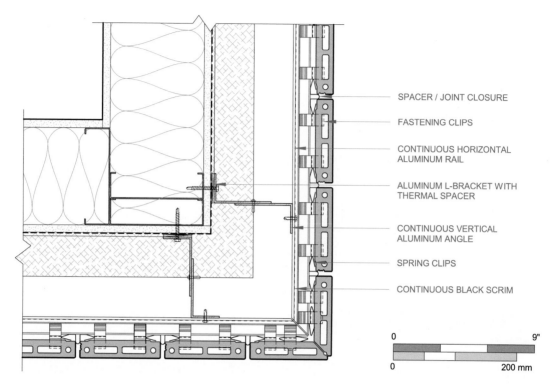

SPACER / JOINT CLOSURE

FASTENING CLIPS

CONTINUOUS HORIZONTAL
ALUMINUM RAIL

ALUMINUM L-BRACKET WITH
THERMAL SPACER

CONTINUOUS VERTICAL
ALUMINUM ANGLE

SPRING CLIPS

CONTINUOUS BLACK SCRIM

0   9"

0   200 mm

Figure 9.5.2 Plan detail at the mitered corner of Tykeson Hall. The rainscreen cladding is attached to steel stud wall construction using vertical and horizontal aluminum rails. Terra cotta by Shildan uses spring clip attachments and closure pieces that block water entry at the vertical joints. From drawings by Rowell Brokaw Architects.

order, color samples, shop drawings and parts inventory. The long lead time meant that choices regarding the terra cotta dominated subsequent thinking about other materials. The strict dimensional control with 3/8 inch (9.5 mm) gaps limited the ability to adjust the tile layout to discrepancies in the window installation (Figure 9.5.6). If the tiles were not perfectly straight, the tight gaps made that obvious and specific lengths and colors limited the opportunity to swap these tiles to the upper floor where they would be less visible. Although the terra cotta required a heavier gauge steel stud wall, it was otherwise easily supported by the post-tensioned concrete slabs. The modest height of the building facilitated rapid hand setting of the tiles from hydraulic lifts (Figure 9.5.7).

The cost of brick veneer on the lower floors was approximately $44.00 per square foot ($473.61/sqm). This was elevated as much as 25% by the custom herringbone pattern and by prevailing wage requirements that eliminated small scale masonry firms from consideration. The terra cotta cost $90.00 per square foot ($968.75/sqm) including $3.50 ($37.67/sqm) for the custom color layout. These numbers compare relatively high-end brick masonry with very efficient use of terra cotta. They also reflect limited

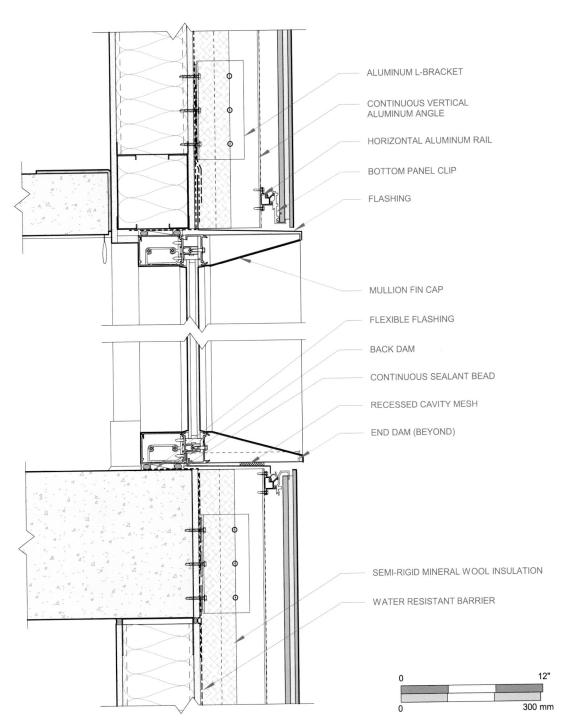

ALUMINUM L-BRACKET

CONTINUOUS VERTICAL
ALUMINUM ANGLE

HORIZONTAL ALUMINUM RAIL

BOTTOM PANEL CLIP

FLASHING

MULLION FIN CAP

FLEXIBLE FLASHING

BACK DAM

CONTINUOUS SEALANT BEAD

RECESSED CAVITY MESH

END DAM (BEYOND)

SEMI-RIGID MINERAL WOOL INSULATION

WATER RESISTANT BARRIER

0			12"
0			300 mm

Figure 9.5.3 Vertical section at a typical window opening, locating the barriers needed to complete a "Perfect Wall" assembly. Custom aluminum mullion caps close the wall cavities at the edges of the terra cotta overlay. From drawings by Rowell Brokaw Architects.

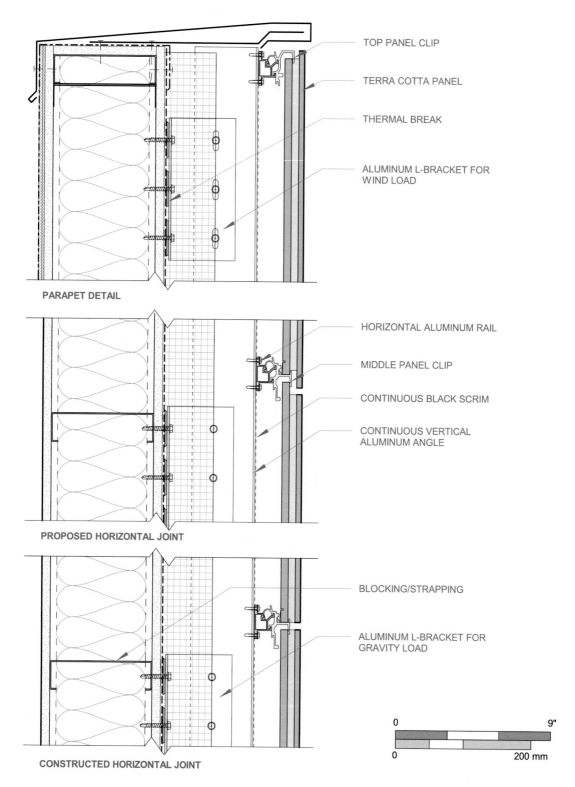

TOP PANEL CLIP

TERRA COTTA PANEL

THERMAL BREAK

ALUMINUM L-BRACKET FOR
WIND LOAD

PARAPET DETAIL

HORIZONTAL ALUMINUM RAIL

MIDDLE PANEL CLIP

CONTINUOUS BLACK SCRIM

CONTINUOUS VERTICAL
ALUMINUM ANGLE

PROPOSED HORIZONTAL JOINT

BLOCKING/STRAPPING

ALUMINUM L-BRACKET FOR
GRAVITY LOAD

CONSTRUCTED HORIZONTAL JOINT

| 0 | | 9" |
| 0 | | 200 mm |

Figure 9.5.4 Vertical section of the wall and parapet assemblies at Tykeson Hall. The original submittal called for overlapping end cuts on the vertically oriented terra cotta. Straight cuts (below) were selected to conceal the clay body color at the horizontal joint lines. From drawings by Rowell Brokaw Architects and Shildan, Inc., Mt. Laurel, New Jersey.

Figure 9.5.5 Left: Comparing the installed glaze colors with the sample tiles that were approved prior to manufacture. Photo: Rowell Brokaw Architects. Right: Field painted spring clips that secure the vertical extrusions. Photo: Donald Corner.

experience with terra cotta in the local market. Regardless, the change of materials was important to the design. Three stories of brick addressed the scale of nearby buildings. A lesser material on the fourth floor and the prominent west façade would have devalued the building in its context. The small amount of terra cotta used raised the unit cost, however it limited the total upcharge for an enduring, high quality finish. At 1% of the total project cost, it was comparable to the required allocation for public art (Figure 9.5.8).

Figure 9.5.6 Full height terra cotta volume at the northwest corner of Tykeson Hall. Window openings are coordinated with the tight layout of vertical tiles. Window transitions on all four edges are accomplished with custom mullion caps. Photo: Frank Visconti for Rowell Brokaw Architects.

Figure 9.5.7 Above: Installation of the vertical aluminum rails, insulation, continuous black scrim and terra cotta rainscreen accomplished from hydraulic lifts. Below: Detail of the wall assembly with the vertical closure and spacing gaskets between the tiles. Photos: Donald Corner.

Figure 9.5.8 The Tykeson Hall glaze colors were derived from building elements in the context, with added highlights. The colors resonate with accents in the brick blend below. The anchored brick veneer is laid in a pattern bond derived from historic campus buildings. Photo: Frank Visconti for Rowell Brokaw Architects.

Bibliography

Allen, Harris, editor. "Terrra Cotta Versus Terra Firma," *California Arts and Architecture*, February, 1930.

American Architect, November 20, 1925: 429.

Architectural Aluminum Manufacturers Association. "The Rain Screen Principle and Pressure Equalized Wall Design," *Aluminum Curtain Walls*. Chicago: Architectural Aluminum Manufacturers Association, February, 1971.

Ault, Nicholas. "Unpacking: A Study in the Generation of Louis Sullivan's Ornament," *2019 Annual Proceedings*. Association of Collegiate Schools of Architecture Press, March 13, 2019.

Bechthold, Martin, Anthony Kane and Nathan King. *Ceramic Material Systems in Architecture and Interior Design*. Basel: Birkhauser Verlag GmbH, 2015.

Brookes, Alan J. and Maaten Meijs. *Cladding of Buildings*. New York: Taylor & Francis, 2008.

Buchanan, Peter. *Renzo Piano Building Workshop: Complete Works, Vol. 3*. London: Phaidon Press Limited, 2000.

Buchanan, Peter. *Renzo Piano Building Workshop: Complete Works, Vol. 4*. London: Phaidon Press Limited, 2003.

Chambers, Cyrus Jr. Brick machine. US Patent # 40,221, October 6, 1863.

Davey, Norman. *A History of Building Materials*. New York: Drake Publishers Ltd., 1971.

Davis, Charles Thomas. *A Practical Treatise on the Manufacture of Bricks, Tiles, Terra-Cotta, Etc.* Philadelphia: Henry Carey Baird & Co., 1884.

Davis, Stanley M. *Future Perfect*. Reading, MA: Addison Wesley, 1987.

Detail: Zeitschrift für Architektur & Baudetail & Einrichtung. "House of Music in Innsbruck," 4 (2019), 54–61.

DSDHA, Architects. *The South Molton Street Building* (London). Inhouse pamphlet prepared for Royal Institute of British Architects awards, October 3, 2013.

Elliot, Cecil D. *Technics and Architecture: The Development of Materials and Systems for Building*. Cambridge, MA: MIT Press, 1992.

Ferriday, Virginia Guest. *The Last of the Handmade Buildings: Glazed Terra Cotta in Downtown Portland*. Portland: Mark Publishing Company, 1984.

Flagge, Ingeborg, Verena Herzog-Loibl and Anna Meseure, editors. *Thomas Herzog: Architecture + Technology.* Deutsches Arkitecturmuseum; Munich, London, New York: Prestel, 2001.

Garófalo, Laura and Omar Kahn, editors. *Architectural Ceramic Assemblies Workshop: Bioclimatic Ceramic Assemblies I* (2016). Hong Kong: Regal Printing, 2017.

Garófalo, Laura and Omar Kahn, editors. *Architectural Ceramic Assemblies Workshop: Bioclimatic Ceramic Assemblies II* (2017). Hong Kong: Regal Printing, 2018.

Garófalo, Laura and Omar Kahn, editors. *Architectural Ceramic Assemblies Workshop: Bioclimatic Ceramic Assemblies III* (2018). Hong Kong: Regal Printing, 2019.

Gartner, Josef, Permasteelisa Group. "Behörde fü Stadtenwicklung und Umwelt, Hamburg, Wihelmsburg (Germany)," Gartner Kalendar 2013. Inhouse promotional publication. https://issuu.com/permasteelisagroup/docs/gartner-kalender-2013?e=5255180/1157970

Geer, Walter. *The Story of Terra Cotta.* New York: Tobias A. Wright Printer and Publisher, 1920.

Glenn Howells Architects. *Design and Access Statement: Lincoln House, Manchester* (UK). Inhouse document in support of planning application for 125 Deansgate, September 16, 2015.

Gruner, Lewis, editor, with illustrations and text by Vitttore Ottolini and Federigo Lose. *The Terra-Cotta Architecture of North Italy (XIIth-XVth Centuries).* London: John Murray, 1867.

Gulling, Dana K. *Manufacturing Architecture: An Architect's Guide to Custom Processes, Materials, and Applications.* London: Lawrence King Publishing Ltd., 2018.

Gwynn, John. *London Improved.* London, 1766.

Hamrick, James Marchall, Jr. *A Survey of the Use of Architectural Terra Cotta in American Commercial Architecture: 1870–1930.* Master of Arts Thesis, Department of Art History, University of Oregon, December 1979.

Haug, Anders, Klaus Rhode Ladeby and Kasper Edwards. "From Engineer-To-Order to Mass Customization," *Management Research News*, 32(7) (2009), 633–44.

Heathcote, Edwin. *Eric Parry Architects*, Vol. 3. London: Artifice Press, 2014.

Henry, Alison, Iain McCraig, Clara Willett, Sophie Godfraind and John Stewart, editors. *Earth, Brick and Terra Cotta.* Farnham: Ashgate, 2015.

Herzog, Thomas. *Thomas Herzog Buildings 1978–1992.* Stuttgart: Verlag Gerd Hatje, 1993.

Herzog + Partner BDA: Thomas Herzog, Hanns Jorg Schrade and Roland Schneider. *Sustainable Height: Deutsche Messe AG Hannover Administration Building.* Munich, London, New York: Prestel, 2000.

Howe, Samuel. "Polychrome Terra Cotta," *American Architect and Building News* (1912), 99–105.

Kurutz, Gary F. *Architectural Terra Cotta of Gladding, McBean*. Sausalito: Windgate Press, 1989.

Ledderose, Lothar. *Ten Thousand Things: Module and Mass Production in Chinese Art*. Princeton: Princeton University Press, 2000.

Leslie, Thomas. *Chicago Skyscrapers, 1871–1934*. Urbana: University of Illinois Press, 2013.

Lstiburek, Joseph. *Insight 001: The Perfect Wall*. Westford: Building Science Corporation, May 2008.

McGinnis, William C. *History of Perth Amboy, N.J., 1651–1959*. Perth Amboy: American Publishing Company, 1959.

National Terra Cotta Society. *Terra Cotta Standard Construction*, Revised Edition. New York: National Terra Cotta Society, 1927.

Norton, F.H. *Elements of Ceramics*. Cambridge, MA: Addison-Wesley Press, Inc., 1952.

Poletti, Therese. *Art Deco San Francisco: The Architecture of Timothy Pflueger*. New York: Princeton Architectural Press, 2008.

Prudon, Theodore Henricus Maria. *Architectural Terra Cotta and Ceramic Veneer in the United States Prior to World War II: A History of Its Development and an Analysis of Its Deterioration Problems and Possible Repair Methodologies*. PhD Dissertation, Columbia University, 1981.

Roth, Leland. *The Urban Architecture of McKim, Mead and White: 1870–1910*. PhD Dissertation, Yale University, 1973.

Sauerbruch Hutton. "Brandhorst Museum." Unpublished press package.

Simpson, Pamela H. *Cheap, Quick & Easy: Imitative Architectural Materials, 1870–1930*. Knoxville: University of Tennessee Press, 1999.

Stanford, Caroline. "Revisiting the Origins of Coade Stone," *The Georgian Group Journal*, XXIV (September 2016).

Stratton, Michael. *The Terra Cotta Revival*. London: Victor Gollancz, a Cassell imprint, 1993.

Sullivan, Louis. *A System of Architectural Ornament According with a Philosophy of Man's Powers*. New York: Press of the American Institute of Architects, Inc., 1924.

Sullivan, Louis. "The Tall Office Building Artistically Reconsidered," *Inland Architect*, XXVII(4) (May 1896).

"Technical Review, The Fisher Building," special supplement, *Inland Architect*, XXVII(4) (May 1896).

Toraldo di Francia, Cristiano. *Art and Technique of the Vertical Terra Cotta Surface: From the Facing to the Vertical Wall*. Florence: Il Palagio srl Cotto Pregiato Imprunetino, 2002.

Tunick, Susan. *Architectural Terra Cotta*. New York: Lumen, Inc., 1986.

Weisman, Winston. "The Commercial Architecture of George B. Post," *Journal of the Society of Architectural Historians*, 31(3) (October 1972), 176–204.

Wells, Jeremy. "The History of Structural Hollow Clay Tile in the United States," *Construction History*, 22 (2007), accessed July 2, 2019.

White, Charles E. Jr. *Architectural Terra Cotta*. Great Britain: International Textbook Company, 1938. Revised and enlarged by Charles Y. Sierks, 1950.

World's Oldest Industry: Fingerprints on the Clays of Time. W.S. Dickey Clay Manufacturing Company, 1924.

SITE VISITS

Boston Valley Terra Cotta, Orchard Park, New York. Plant tour. August 16, 2019.

Gladding McBean, Lincoln California. Plant tours with Jamie Franham, National Sales Manager, December 16, 2014 and December 18, 2018.

Guaranty Building Interpretive Center, Buffalo, New York. Site visit. August 15, 2019.

Museo degli Innocenti, Florence, Italy. Site visit. April 16, 2019.

NBK Keramik, Emerich, Germany. Plant tour with Stefan Verriet, Sales, April 4, 2019.

Palagio Engineering, Greve in Chianti (Florence), Italy. Plant tours with Alexander Piazza, Export Manager, April 30, 2012 and June 21, 2019.

INTERVIEWS

Alesch, Stephen, Principal, Roman and Williams. Telephone interview with Donald Corner, September 21, 2020.

Carpenter, James, Principal, James Carpenter Design Associates. Telephone interview with Donald Corner, February 25, 2020.

Ceder, Erica, Architect DLR Group, Portland, Oregon. Interview with Donald Corner, March 6, 2019.

Crosby, Nat, Architect. Telephone interview with Donald Corner, March 16, 2019.

Desborough, Adam and Sandeep Shambi, Project architect and Director, Glenn Howells Architects. Electronic meeting with Donald Corner, April 21, 2020.

Devine, Antonia, JDS Development. Telephone interview with Donald Corner, August 7, 2020.

Farnham, Jamie, National Sales Manager, Gladding McBean. Electronic meeting with Donald Corner, January 29, 2021.

Friedman, John, JFAK Architects. Telephone interview with Donald Corner, March 5, 2020.

Fritz, Michael, Sculptor, Boston Valley Terra Cotta. Zoom interview with Donald Corner, January 14, 2021.

Lehmann, Christian, General Manager, NBK North America, Salem, Massachusetts. Telephone interview with Donald Corner, March 25, 2019.

Lehmann, Christian, General Manager, NBK North America, Salem, Massachusetts. Electronic conference with Donald Corner, November 10, 2020.

Olson, Paul, Principal, Olson Kundig Architects. Telephone interview with Donald Corner, January 20, 2020.

Richardson, Simon, FCB Studios. Electronic conference with Donald Corner and John Rowell, July 31, 2020.

Streff, F.J. (Bud), Director of Sales, NBK North America, Salem, Massachusetts. Telephone interview with Donald Corner, March 25, 2019.

Van der Staay, Maurits, Architect. Sequence of electronic conferences with Donald Corner for the development of case study materials, April 1, 2020 through September 23, 2020.

WEB REFERENCES

ASTM International. "ASTM C67/C67M-20: Standard Test Methods for Sampling and Testing Brick and Structural Clay Tile." www.astm.org/Standards/C67.htm, accessed March 23, 2021.

Blue Mountains Creative Arts Center. "Learn Glaze Chemistry in 15 Minutes," Saturday Potters Glaze Workshop (August 24, 2019). www.youtube.com/watch?v=BT1s48h8P20, accessed May 13, 2020.

Boston Valley Terra Cotta. "Connecticut College, New London Hall Life Science Building," Portfolio. https://bostonvalley.com/portfolio-item/connecticut-college-new-london-hall-life-science-building/, accessed September 5, 2019.

Boston Valley Terra Cotta. "The Fitzroy in NYC Exemplifies Boston Valley's Thorough QA/QC Process Demonstrated with this Large Dry-fit Mockup." https://bostonvalley.com/514-west-24th-the-fitzroy-dry-fit/, accessed August 7, 2020.

Chen, Michael K., Michael K. Chen, Architects. Online Portfolio. http://mkca.com/projects/upper-east-side-townhouse/, accessed February 11, 2019.

Collins, Glenn. "9/11's Miracle Survivor Sheds Bandages; A 1907 Landmark Will Be Restored for Residential Use." *New York Times*, March 5, 2004. www.nytimes.com/2004/03/05/nyregion/9-11-s-miracle-survivor-sheds-bandages-1907-landmark-will-be-restored-for.html

CookFox, Architects. Online Portfolio. https://cookfox.com/projects/512-west-22nd-street/, accessed February 11, 2019.

Digital Fire Corporation. "Creating a Non-glaze Ceramic Slip or Engobe," Digital Fire: Engobes. https://digitalfire.com/article/creating+a+non-glaze+ceramic+slip+or+engobe, accessed May 15, 2020.

Digital Fire Corporation. "Glaze Chemistry." https://digitalfire.com/glossary/glaze+chemistry, accessed May 13, 2020.

Digital Fire Corporation. "Matte Glaze." https://digitalfire.com/glossary/matte+glaze, accessed May 16, 2020.

Hall, John F. "Temples, Tombs, and Etruscan Treasures: From Tuscany to Dallas." *American Journal of Archaeology* 113(3) (2009), 463–9. http://www.jstor.org/stable/20627598, accessed March 24, 2021.

Jones, Craig and Geoffrey Hammond, Bath University. "Inventory of Carbon and Energy (ICE) V.3.0," ICE. https://circularecology.com/embodied-carbon-footprint-database.html, accessed November 23, 2020.

NBK Keramik GmbH. "BSU Hamburg." https://nbkterracotta.com/en/project/bsu-hamburg/, accessed November 24, 2020

NBK Keramik GmbH. "NBK M-9 Mestre." Press release.

New York City School Construction Authority. "Standard Specifications: 04250 Terra Cotta." www.nycsca.org/Design/Design-Standards#Specifications-86, accessed January 31, 2021.

SHoP Architects. "Wave/Cave." www.shoparc.com/projects/wave-cave, accessed February 20, 2020.

Terreal North America. "History & Success." www.terrealna.com/the-terreal-difference/history-and-success/, accessed July 31, 2020.

Zoning Resolution of the City of New York. "Permitted Obstructions in Certain Districts," Article II, Chapter 3, Section 23–621. https://zr.planning.nyc.gov/article-ii/chapter-3#23-621, accessed September 29, 2020.

Index

Note: *Italic* page numbers refer to figures